ARIZONA LODE GOLD MINES AND GOLD MINING

by

ELDRED D. WILSON, J. B. CUNNINGHAM, AND G. M. BUTLER

THE ARIZONA BUREAU OF MINES

BULLETIN 137

Revised 1967

THE UNIVERSITY OF ARIZONA

TUCSON

FOREWORD

Arizona Bureau of Mines Bulletin No. 137, ARIZONA LODE GOLD MINES AND MINING, was originally issued in 1934 in response to the heavy demand that was being expressed during the early 1930's for authoritative information on Arizona's gold deposits. After several reprintings, the bulletin was allowed to go out of print during the early years of World War II and, in view of the restrictions placed on gold mining during the war, it was not considered necessary to reprint the bulletin at that time. Circumstances now have changed, however, and interest in gold is again strong and inquiries are being received almost daily for information on gold deposits.

With this in mind and considering that the data on the geologic setting and the mineralogical characteristics of the deposits described in Bulletin No. 137 are as valid today as they were in 1934, it was decided to reprint Parts I and II of the Bulletin without change. Much of Part III, which deals with the laws and regulations governing mining, however, have been deleted, inasmuch as many of the laws have been revised since 1934. For information on this aspect of mining, the reader is referred to any of the recently issued books and pamphlets on mining law.

We are pleased to acknowledge the financial assistance given by the U.S. Department of Commerce through its State Technical Services Program in the reprinting of this bulletin.

J. D. Forrester, Director
Arizona Bureau of Mines

PREFACE

In 1929, thirty-nine mines or prospects in which gold was the
element principally sought were listed by the State Mine In-
spector of which four actually produced some gold. In his report
for 1933, however, the official mentioned lists seventy-four such
properties and over half of them produced more or less gold
during the year. These facts prove that the depression, which has
thrown many men out of work, and the great increase in the
price of gold have created an interest in prospecting for and
mining gold such as has not existed for many years.

Over twenty-six hundred persons are now prospecting for
placer or lode deposits of gold or operating on such deposits in
Arizona, whereas less than four hundred persons were employed
in gold mines in the State in 1929 and only a few score prospectors for gold were then in the field. This tremendously increased interest in gold has created a great demand for information about Arizona's gold resources, which this bulletin is intended to supply. Because Arizona has for many years produced
more copper than any other state in the Union, it is not generally
recognized that gold ores are widely distributed throughout the
State, and it is the hope of the authors that the publication of the
facts will attract more capital to Arizona for the development of
her gold deposits. It is also hoped that the sections on working
small gold lodes, on the laws and regulations relating to the location and retention of lode claims, and on prospecting for gold
will be helpful to inexperienced prospectors.

This bulletin completes a series of three on gold that have been
published by the Arizona Bureau of Mines, the other two being
No. 133 (Treating Gold Ores) and No. 135 (Arizona Gold Placers
and Placering). These three bulletins contain practically all the
general information available on the subjects covered by them.
Subsequent bulletins that relate to gold mining in Arizona will
take the form of detailed reports on various mining districts.

It has been impossible for the Bureau with its limited staff and
funds to make this publication complete. It is quite probable that
a number of meritorious properties are not mentioned herein.
The fact that no reference is found in this publication to some
property should not be considered as proof, or even an indication,
that the property is unworthy. Moreover, it should not be assumed that all the properties described are being managed effi-
ciently and honestly and that their operations are sure to be
profitable. It is not the duty of the Arizona Bureau of Mines to
investigate the management of mining companies or to advise
people concerning investment in mining stocks.

G. M. BUTLER.

July 1, 1934.

TABLE OF CONTENTS

PART I

ARIZONA LODE GOLD MINES

9

PART II

THE OPERATIONS AT SMALL GOLD MINES

PART III

LAWS, REGULATIONS, AND COURT DECISIONS BEARING ON THE LOCATION AND RETENTION OF LODE CLAIMS, TUNNEL SITES, AND MILL SITES IN ARIZONA

PART IV

SOME HINTS ON PROSPECTING FOR GOLD

LIST OF ILLUSTRATIONS

PAGE

PLATES

PART I

ARIZONA LODE GOLD MINES

By Eldred D. Wilson

CHAPTER I—INTRODUCTION

DISTRIBUTION, GEOLOGIC SETTING, AND TYPES OF DEPOSITS

Of the approximately $2,800,000,000 worth of minerals that have been produced by Arizona, gold has constituted about 5.6 per cent, or some $158,000,000 (see page 18). Of this gold production, more than $84,000,000 worth or 53 per cent has come from lode gold and silver mines, nearly 40 per cent as a by-product from copper, lead, and zinc mining, and 7 per cent from placers.

Figure 1 shows the distribution of lode gold deposits in Arizona. No commercial gold deposits have been found in the northeastern or Plateau portion, which consists of a thick, relatively undisturbed succession of Paleozoic, Mesozoic, and Tertiary rocks resting upon a basement of pre-Cambrian schist, gneiss, and granite. Nearly 80 per cent of the State's lode gold production has come from deposits that occur within a distance of 65 miles from the southwestern margin of the Plateau. These deposits occur within the belt that, in Arizona, Nevada, Utah, Colorado, and New Mexico, borders the Colorado Plateau and, as B. S. Butler points out, contains more than 75 per cent of the mineral deposits of the Southern Rocky Mountain region. This belt has been subjected to intense faulting and igneous intrusion. Notable gold deposits occur also southwest of this belt, in areas of intense faulting and intrusion by stocks.[1]

---

[1] Butler, B. S., Ore deposits as related to stratigraphic, structural, and igneous geology in the western United States: Am. Inst. Min. Eng., Lindgren Volume, pp. 198-240, 1933; Relation of ore deposits of the Southern Rocky Mountain region to the Colorado Plateau: Colo. Sci. Soc., Bull. 12, pp. 30-33, 1930.

Figure 1.—Index map showing location of lode gold districts in Arizona.

1 Lost Basin
2 Gold Basin
3 Northern Black Mountains (Weaver, Pilgrim)
4 Union Pass
5 Oatman
6 Musie Mountain
7 Cerbat Mountains (Wallapai)
8 McConnico
9 Maynard
10 Cottonwood
11 Chemehuevis
12 Cienega
13 Planet
14 Plomosa
15 La Paz
16 Ellsworth
17 Kofa
18 Sheep Tanks
19 Tank Mountains
20 Gila Bend Mountains
21 Trigo Mountains
22 Castle Dome
23 Las Flores (Laguna)
24 La Posa

25 La Fortuna
26 Eureka
27 Prescott, Groom Creek
28 Cherry Creek
29 Squaw Peak
30 Hassayampa, Walker, Big-bug, Turkey Creek
31 Black Canyon
32 Peck, Bradshaw, Pine Grove, Tiger, Minnehaha
33 Humbug, Castle Creek
34 Black Rock, White Picacho
35 Weaver (Octave)
36 Martinez
37 Vulture
38 Big Horn
39 Midway
40 Agua Fria
41 Cave Creek
42 Winifred
42-a Salt River
43 Payson (Green Valley)
44 Spring Creek
45 Globe
46 Banner or Dripping Springs

47 Goldfields
48 Superior (Pioneer)
49 Saddle Mountain
50 Cottonwood
51 Mammoth (Old Hat)
52 Casa Grande
53 Owl Head
54 Old Hat
55 Quijotoa
56 Puerto Blanco Mountains
57 Comobabi
58 Baboquivari
59 Greaterville
60 Arivaca
61 Oro Blanco
62 Wrightson
63 Gold Gulch (Morenci)
64 Twin Peaks
65 Lone Star
66 Clark
67 Rattlesnake
68 Dos Cabezas
69 Golden Rule
70 Tombstone
71 Turquoise
72 Huachuca

The lode gold deposits in Arizona are of Lindgren's epithermal, mesothermal, and hypothermal types.[2] The epithermal veins have furnished nearly half of the production, the mesothermal slightly more than half, and the hypothermal less than one per cent.

Epithermal veins: Representative of the epithermal type are the veins of the Oatman, Union Pass, Kofa, and Sheep Tanks districts. These veins were deposited under conditions of moderately low pressure, at depths generally less than 3,000 feet below what was then the surface. They are best developed in areas of Tertiary volcanic activity and are of Tertiary age. The veins are characterized by rather irregular form, with rich ore bodies relatively near the surface, only; fine-grained to chalcedonic greenish-yellow quartz, commonly with microscopic adularia; marked banding or crustification due to successive stages of deposition; breccia inclusions; and wall-rock alteration to chlorite, carbonates, quartz, and fine-grained sericite. Calcite is commonly abundant in the gangue, and the quartz locally shows lamellar structure pseudomorphic after calcite. Their gold generally occurs as very finely divided, pale-yellow particles, alloyed with more or less silver. Gold and silver tellurides, generally so common in epithermal deposits, have not been found in Arizona.

The epithermal veins have formed no placers of economic importance.

Mesothermal veins: Representative of the mesothermal type are most of the veins of the Bradshaw, Weaver, Date Creek (Congress), Vulture, Harquahala, Gila (Fortuna), and Dos Cabezas Mountains. These veins were deposited under conditions of moderately high temperature and pressure, at depths of more than 3,000 feet below what was then the surface. They occur in schist, gneiss, granite, and sedimentary rocks, and are predominantly of Mesozoic or Tertiary age. In general, they are persistent and are characterized by rather regular form; localization by fractures with even to smooth walls; coarse-grained texture; banding due mainly to shearing and replacement; and wall-rock alteration to carbonates, quartz, and rather coarse-grained sericite. They generally contain sulphides, particularly pyrite, chalcopyrite, galena, sphalerite, and arsenopyrite. Their gold occurs both as fine to coarse free particles and as a finely divided constituent of the sulphides. Many of them contain more silver than gold by weight. Most of them have not been of commercial grade below depths of several hundred feet, but the Congress and Octave veins were mined to depths of 4,000 and 2,000 on the dip, respectively.

Where veins of this type carried fairly coarse free gold, placers have been formed.

Hypothermal veins: Hypothermal veins of economic importance are not abundant in Arizona. They are represented in the

---

[2] Lindgren, Waldemar, Mineral deposits, 4th. ed., 1933.

Cherry Creek district and in portions of the Bradshaw Mountains area, of Yavapai County. Their characteristics are summarized on page 22.

HISTORY OF ARIZONA GOLD MINING[1]

Gold mining in Arizona did not start to any appreciable extent until after the acquisition of the territory by the United States from Mexico in 1848 and 1853. What little mining was done by the Spanish and Mexican miners was for silver. A little placer gold was brought in to the churches by Indian converts from the dry working of gravels in the desert, but no systematic mining was done.

After the final occupation of Arizona in 1853, the only accessible part of the Territory was that around the old Mexican settlements of Tucson and Tubac. Considerable prospecting was done in this part of the Territory by American prospectors, and several silver mines and one copper mine were opened, but little or no gold mining was done. On the outbreak of the Civil War, the withdrawal of troops opened the door to Apache raids, and all mining ceased.

During the Civil War, prospectors entered the Territory with the California troops, and several exploring parties were organized to hunt for gold in the central part of the State, hitherto an unknown wilderness dominated by Apaches. Rich placers were found near the Colorado River at Gila City, La Paz, and Quartzsite, and soon after the Rich Hill, Lynx Creek, Hassayampa, and Big Bug placers in the Bradshaw Mountains of central Arizona were discovered. Base metal mines and even silver mines were not sought, as only gold could be mined at a profit from this inaccessible and hazardous corner of the world. After the richer parts of the placers were exhausted, gold ledges were located and worked in the crudest manner. Most of the free-milling ore proved superficial. Only one large deposit, the Vulture, was exploited on a large scale.

At the end of the Civil War, troops were again withdrawn, resulting in ten years of chaos and bloody warfare with the Apaches. Little mining was done except around Prescott and Wickenburg where some protection was given by troops guarding Prescott, then the capital of the Territory.

Finally, in 1872, large reservations were set aside for the Indians and the first truce was declared. The country was then enjoying the post-Civil War period of high commodity prices. Gold was relatively low in price as compared with silver and copper. Prospecting for these two metals, on the establishment of peace with the Indians, took precedence over gold, resulting in the succeeding ten years, in the discovery and exploitation of rich silver mines in the Bradshaws, Silver King, Signal, Globe, and

---

[1] By J. B. Tenney.

Tombstone. This silver boom was followed after the completion of the two transcontinental railroads in 1881 by the discovery and early exploitation of nearly every copper deposit in the Territory.

From 1884 to 1893 the country went through a severe deflation of commodity values. The copper and silver markets fell rapidly resulting in a relative rise in the price of gold. On the demonitization of silver in 1893, practically all silver mining ceased, and only the richest and largest copper mines continued to operate.

From 1893 to 1900, miners from all the old silver camps of the West again turned to the search for gold, which resulted in Arizona in the discovery of numerous new gold deposits, more notably the Congress and Octave in the Bradshaw Mountains, the Mammoth north of Tucson, and the rich Harqua Hala, La Fortuna, and King of Arizona mines in the desert of Yuma County. The development of the cyanide process and of better concentration methods encouraged the reopening of numerous old mines near Prescott and the exploitation of the deeper base ore.

Towards the end of the nineteenth century, the long period of stagnation ended and commodity prices again turned upwards. Gold mining became less attractive, and the miners in Arizona turned their attention to copper. From 1900 until the business collapse of 1929 and 1930, gold mining was subordinate to base-metal mining. The only exceptions were the discovery and exploitation of the rich vein deposits of the Gold Road, Tom Reed, United Eastern, and others, in the Oatman district. Gold mining also continued on a reduced scale in the older mines of the Bradshaw Mountains and in those of Yuma County.

On the collapse of commodity prices in 1930, miners again turned their attention to gold. The first result was the search for new placers and the reworking of old fields, with indifferent results. The higher gold prices that were established by the United States in 1933 have revived activity in most of the old gold camps and stimulated prospecting throughout the State. In 1933, production was about 12 per cent greater than in 1932.

Arizona has produced more non-ferrous metallic wealth than any state or territory in the Union. While most of this production has been in copper, nearly every copper mining operation in the State has yielded important quantities of gold.

As a gold producer, Arizona ranks seventh in the United States. In the following table, the Arizona gold production is shown segregated as to its source. As is seen, about 40 per cent has come as a by-product of copper and lead mining, chiefly after 1900.

GOLD PRODUCTION OF ARIZONA

(Values Based on a Price of $20.67 Per Ounce of Fine Gold)

PERIOD	From Lode Gold Mines		From Placers		From Copper Mines		From Lead Mines		TOTAL
	Value	Per cent	Value	Per cent	Value	Per cent	Value	Per cent	Value
1853-1872, incl.	Approx. $ 2,880,000	36.4	Approx. $ 5,000,000	63.6					Approx. $ 7,860,000
1873-1893, incl.	Approx. $ 9,500,000	73.5	Approx. $ 3,420,000	26.5					Approx. $ 12,920,000
1898-1903, incl.	Approx. $16,907,000	70.1	Approx. $ 1,800,000	7.5	Approx. $ 4,893,000	20.3	Approx $ 500,000	2.1	Approx. $ 24,100,000
1904-1930, incl.	$53,387,000	49.6	$ 518,000	.5	$49,911,000	46.4	$3,744,000	3.5	$107,560,000
1931-1933, incl.	Approx. (a) $ 1,706,000	(a) 30.7	(a)	(a)	Approx. (b) $ 3,854,000	(b) 69.3	(b)	(b)	Approx. $ 5,560,000
Grand Total 1853-1933, incl.	Approx. $84,360,000	53.4	Approx. $10,738,000	6.8	Approx. $58,658,000	37.1	Approx. $4,244,000	2.7	Approx. $158,000,000

(a) Gold lode and placer mines combined.
(b) Copper and lead mines combined.

CHAPTER II—YAVAPAI COUNTY

TOPOGRAPHY

As show on Figure 2, Yavapai County comprises an irregular area about 105 miles long from west to east by 104 miles wide. Except for the margin of the Plateau Province along its northeastern boundary, this area consists of a series of northward- and northwestward-trending fault-block mountain ranges and valleys. The largest of these ranges is the Bradshaw, which is about 45 miles long by 20 miles wide and attains a maximum altitude of 7,971 feet above sea level. Farther west, the mountains are somewhat smaller but equally rugged. As shown on Figure 2, the region is drained by the Verde, Agua Fria, Hassayampa, and Santa Maria rivers. The lowest point, about 1,500 feet above sea level, is on the Santa Maria River.

CLIMATE AND VEGETATION

The climatic range of Yavapai County is illustrated by the following examples:

The highest and lowest temperatures on record for Prescott (altitude 5,320 feet) are 105° and —12°, while for Clemenceau (altitude 3,460 feet) they are 110° and 13°, respectively. The normal annual precipitation at Crown King (altitude 6,000 feet) is 32.42 inches, while at Prescott it is 18.52 inches. This higher country is subject to considerable snow in winter. The lower portions of the region receive only from 10 to 13 inches of rainfall yearly.[4]

In general, the higher ridges and valleys are well wooded and watered, while the slopes below 5,000 feet in altitude are brushy, and the country below 3,500 feet is characterized by desert vegetation.

ROUTES OF ACCESS AND TRANSPORTATION

As shown by Figure 2, the Phoenix line of the Santa Fe Railway crosses Yavapai County from north to south. A branch from this line connects Clemenceau, Clarkdale, and Jerome with the Phoenix line. Another branch via Mayer and Humboldt, to Crown King was partly dismantled in 1926.

Various highways, improved roads, and secondary roads traverse the county and lead from Prescott, the principal distributing center, to the mining districts (see Figure 2).

Since the closing of the Humboldt smelter, the gold smelting ores and concentrates have generally been sent to El Paso and Superior.

---

[4] Smith, H. V. The climate of Arizona: Univ. of Ariz. Agri. Exp. Sta., Bull. 130, 1930.

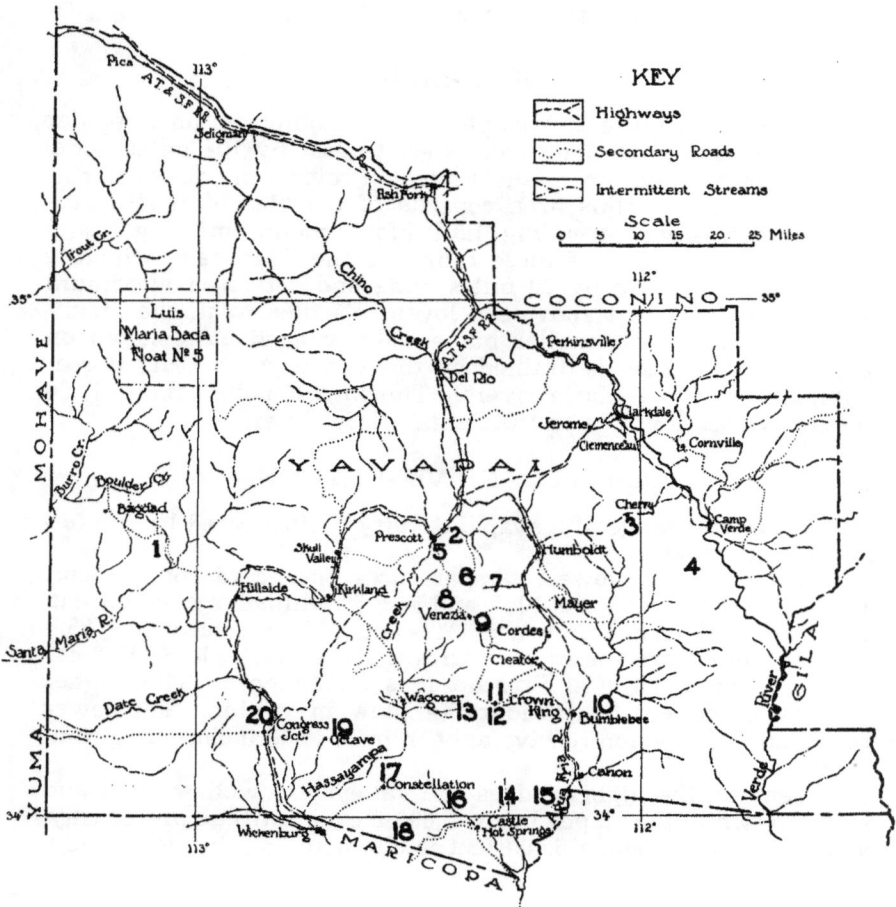

Figure 2.—Map showing location of lode gold districts in Yavapai County.

1	Eureka	11	Peck, Bradshaw
2	Prescott	12	Pine Grove, Tiger
3	Cherry Creek	13	Minnehaha
4	Squaw Peak	14	Humbug
5	Groom Creek	15	Tip Top
6	Walker	16	Castle Creek
7	Bigbug	17	Black Rock
8	Hassayampa	18	White Picacho
9	Turkey Creek	19	Weaver
10	Black Canyon	20	Martinez

GENERAL GEOLOGY

Except for certain areas of Paleozoic sedimentary beds in the northern and northeastern portions of the county, the mountains are made up of metamorphic and igneous rocks.

The oldest formation, the Yavapai schist[5], consists of metamorphosed pre-Cambrian sedimentary and igneous rocks which have been crumpled into generally northeastward-trending belts, cut by various intrusives, and subjected to complex faulting.

The principal intrusives consist of dikes and stocks of diorite, batholithic masses of granite with pegmatites, stocks of granodiorite and monzonite-porphyry, and dikes of rhyolite-porphyry. The diorite and granite are of pre-Cambrian age, while the granodiorite and monzonite-porphyry are, because of their composition, regarded as Mesozoic or early Tertiary. The rhyolite-porphyry dikes cut the granodiorite and are also probably Mesozoic or early Tertiary.

Tertiary and Quaternary volcanic and sedimentary formations in places mantle large areas of the older rocks.

GOLD DEPOSITS

Production: Yavapai County ranks first among the gold-producing counties of Arizona. It has yielded approximately $50,000,000 worth of gold of which about 50.5 per cent was derived from copper mines, 39 per cent from gold and silver veins, 10 per cent from placers, and 0.5 per cent from lead and zinc mines.[6]

Most of the gold obtained as a by-product of copper mining has come from the Jerome district, where a large amount of the ores have yielded from 0.025 to 0.225 ounces of gold per ton.

Distribution: As indicated by Figure 2, the gold districts are confined to the southern half of the county where faulting and igneous intrusion have been relatively intense.

Types: The principal types of gold deposits in Yavapai County are exemplified in the Bradshaw Mountains and Jerome quadrangles where Lindgren[7] has classified them as follows: (1) Mesozoic or early Tertiary gold and gold-silver veins; (2) pre-Cambrian gold-quartz veins; and (3) pre-Cambrian gold-quartz-tourmaline replacement deposits.

---

[5] Jaggar, T. A. and Palache, C., Bradshaw Mountains Folio: U. S. Geol. Survey, Folio No. 126, 1925.

Lindgren, W., Ore deposits of the Jerome and Bradshaw Mountains quadrangles, Arizona: U. S. Geol. Survey Bull. 782, 1926.

Reber, L. E., Geology and ore deposits of the Jerome district: Am. Inst. Min. and Met. Eng., Trans., vol. 66, pp. 3-26, 1922.

Lausen, Carl, Pre-Cambrian greenstone complex of the Jerome quadrangle: Jour. Geol., vol. 38, pp. 174-83, 460-65, 1930.

[6] Figures compiled by J. B. Tenney.

[7] Work cited, pp. 36-48.

As these types of deposits differ markedly in distribution, mineralogy, texture, and form, their distinction is of primary economic importance. The pre-Cambrian gold deposits of the county have not produced over $1,000,000, whereas the Mesozoic or early Tertiary veins have yielded more than $18,000,000 worth of gold.

Mesozoic or early Tertiary veins: Representative of this type are the Congress, Octave, Humbug, Pine Grove or Crown King, and Walker veins, and most of the deposits in the Eureka, Hassayampa, Groom Creek, Bigbug, and Black Canyon districts.

In general, these veins, though locally lenticular, are persistent, straight, and narrow, with definite walls. Their gangue is massive to drusy milky-white quartz, with or without ankeritic carbonates. Their ore shoots, below the zone of oxidation, contain abundant sulphides, principally pyrite, galena, sphalerite, chalcopyrite, arsenopyrite, and tetrahedrite. Most of them are silver bearing, and many contain more silver than gold by weight. In the primary zone, some of their gold is free, but most of it occurs as sub-microscopic intergrowths with the sulphides, particularly the finer-grained galena, chalcopyrite, and pyrite. Their oxidized zone was rich in free gold but generally shallow. They have yielded placers only where their primary zone carried considerable free gold. The vein wall rocks generally show sericitization and carbonatization. High-temperature minerals are absent.

These veins belong to the mesothermal class and were deposited at depths of 3,000 or 4,000 feet below what was then the surface. They occur in schist, granite, and granodiorite or quartz diorite. As they cut the granodiorite and, in the Bradshaw Mountains, appear to be genetically related to the rhyolite-porphyry dikes, Lindgren regards them as of Cretaceous or early Tertiary age.

Pre-Cambrian veins: Representative of this type are the Cherry Creek, Bullwhacker, Jersey Lily, Ruth, Richinbar, Yellow Aster or Lehman, and Minnehaha deposits.

These veins are characteristically lenticular in form. Their gangue is shiny milky-white to glassy quartz with tourmaline and minor amounts of ankeritic carbonates. Below the zone of oxidation, pyrite, chacopyrite, sphalerite, and galena are locally present but generally not abundant. In the primary zone, the gold occurs both as coarse free particles and as sub-microscopic intergrowths with the sulphides. Some silver is present with the gold. The vein walls show slight alteration. Most of the placers of this region have been derived from veins of this type, but, where a large proportion of the gold occurs in the sulphides, placers are lacking.

These veins belong to the hypothermal class and were deposited under conditions of high temperature and pressure. Lindgren has pointed out that they are probably genetically connected with the pre-Cambrian Bradshaw granite.

Pre-Cambrian gold-quartz-tourmaline replacement deposits:
The only deposit of this type known in the county is that of the
Iron King mine, described on page 36. It is believed to have
been formed by emanations from the pre-Cambrian granite.

HISTORY OF GOLD MINING IN YAVAPAI COUNTY[8]

Prior to the Mexican War, early trail-makers and trappers penetrated the Bradshaw Mountains and reported the presence of
minerals there, but the inaccessibility of the country and the
danger from the Indians discouraged prospecting. It was not
until the Civil War, when troops from California, many of whom
were gold miners, came in, that parties were organized to prospect the area.

In 1862, a party headed by A. H. Peeples and guided by Pauline
Weaver discovered the Rich Hill gold placers. During the following year, the Joseph Walker party, of Colorado, journeyed
through the Hassayampa region and found gold placers on Lynx,
Big Bug, Hassayampa, and Groom creeks.

The discovery in the same year of the rich Vulture ledge, southwest of Antelope Peak, stimulated lode gold prospecting. Many
deposits were found, and their free-milling ores were treated in
crude reduction works, chiefly arrastres and small stamp mills.
Base ore was generally encountered at depths of 50 to 100 feet.

From the discovery of several silver bonanzas, during the early
seventies, until the building of the railway into Prescott, in 1888,
lode gold mining in Yavapai County was subordinate to silver
mining.

During the late eighties, after the exhaustion of the silver mines,
many old gold mines were reopened and new ones, notably the
Congress, Hillside, and Octave, were discovered. During the
early nineties, concentrators to treat base gold ores were erected
at the Senator, Crown King, Little Jessie, and other mines, and a
copper smelter was built at Arizona City.

The perfection of the cyanide process, in the early twentieth
century, was an added stimulus to base gold ore mining. The
most important producers of this period were the Congress, Octave, and McCabe-Gladstone. From 1913 until 1930, little lode
gold mining was done in the county.

EUREKA DISTRICT

The Eureka district, of southwestern Yavapai County, is bound-
ed by the Santa Maria River on the south, Burro Creek on the
northwest, the Mohave County line on the west, and the Phoenix
branch of the Santa Fe Railway on the east. From Hillside, its
principal railway shipping point, the area is accessible by improved roads that lead to Bagdad and Kingman, and by numerous
unimproved routes.

---

[8] By J. B Tenney.

West and northwest of Hillside are the McCloud Mountains, a granite range that attains a maximum altitude of 4,900 feet. Westward, these mountains give way to the ruggedly dissected basin of the Santa Maria and its tributaries, Burro, Boulder, Yavapai, and Bridle creeks. This basin, which descends to a minimum altitude of 1,600 feet, is characterized by desert vegetation and hot summers. Aside from the larger streams, water is obtained from a few springs and shallow wells.

This region is composed mainly of granite, gneiss, and schist, intruded by various dikes and stocks and overlain in places by mesa-forming lavas. It contains the Bagdad copper deposit, the Copper King zinc deposit, and several gold-bearing veins of which the Hillside and Crosby have been notable producers. Although these veins are of the mesothermal quartz-pyrite-galena type, they have been mined mainly in the oxidized zone. Early in 1934, preparations were being made to mine and concentrate ores from the sulphide zone.

HILLSIDE MINE

Situation: The Hillside property of seven claims is in Secs. 16 and 21, T. 15 N., R. 9 W., in the vicinity of Boulder Creek. From the railway at Hillside station, it is accessible by 32 miles of road.

History[9]*:* This deposit was discovered in 1887 by John Lawler. The first ore, which consisted of lead carbonate rich in gold and silver, was shipped to Pueblo, Colorado. From 1887 to 1892, Lawler did about 7,000 feet of development work, built an 84-mile road to Seligman, erected a small stamp mill, and made a considerable production. In 1892, the property was sold to H. A. Warner who organized the Seven Stars Gold Mining Company. Guided in part by the advice of T. A. Rickard, this company carried out considerable development work, erected a mill, and built a road to Hillside station. The Warner companies, however, failed in the 1893 depression, and the Hillside property, after protracted litigation, reverted to Lawler. For several years after 1904, the mine was worked intermittently, mainly by lessees. Upon the death of Lawler in 1917, operations ceased until early in 1934 when H. L. Williams purchased the property and constructed a new 125-ton mill. Regular operations, employing eight men underground and twelve in the mill, began in July, 1934.

Production: Acording to records and estimates by Homer R. Wood, who was engineer at the property during part of its activity, the Hillside mine produced 13,094 tons of ore which yielded 9,329 ounces of gold and 219,918 ounces of silver, in all worth about $296,500.

Topography and geology: The Hillside mine is on the east side of Boulder Creek, in a deep canyon of moderately fissile grayish mica schist. This schist is intruded on the east by coarse-grained

---

[9]Historical data largely from Homer R. Wood, of Prescott.

granite and, 2 miles south of the mine, by the Bagdad granite porphyry stock. Narrow dikes of pegmatite are locally present.

Vein and workings: The Hillside vein occurs within a nearly vertical fault zone that strikes N.-N. 15° E. and, in places, separates into branches a few feet apart. In the vicinity of the vein, the schist dips almost flat.

The vein has been opened by over 10,000 feet of workings, distributed over a length of approximately 2,400 feet. Its width ranges from a few inches to several feet and averages about 1½ feet. On the fourth or deepest level, which is from 60 to 300 feet below the surface, it consists of stringers and irregular bunches of coarse-grained massive white quartz with abundant sulphides.

Microscopic examination of a polished section of ore from the fourth level shows that the sulphides are mainly pyrite and galena, together with some sphalerite. Considerable oxide and carbonate material, containing principally iron, lead, and zinc, is visible. The pyrite ranges from massive texture to grains less than 200-mesh in size. The galena occurs as irregular bodies many of which are visible to the unaided eye. They commonly terminate in small veinlets less than 0.0004 inch wide. The gold occurs both in the oxidized material and in the sulphides, but is most abundant in the galena. The silver of the sulphide zone occurs mainly with the sphalerite. Specimens of wire silver have been found in vugs in all the levels, and cerargyrite is locally abundant in the upper workings.

The vein contains some sulphides on the third level and is practically all oxidized above the second level. It has been largely stoped out from the second level to the surface. The wall rock shows strong alteration to quartz and sericite.

COMSTOCK AND DEXTER MINE

The Comstock and Dexter property is on a tributary of Boulder Creek, about 1½ miles south of the Hillside mine. During the late eighties and early nineties, the Dillon brothers worked this deposit and treated their ore in a small stamp mill. Later, the mine was acquired by John Lawler and associates. According to local people, the deposit produced several thousand dollars' worth of gold.

In 1932, the General Minerals Company obtained the property, built a new camp and a road that connects with the Hillside mine road, and started sinking a new shaft. Operations were suspended in 1933.

Here, mediumly massive gray schist strikes northward and dips almost vertically. The vein strikes and dips essentially with the lamination of the schist. It was opened by more than 500 feet of drifts on two adit levels, and by a new 120-foot vertical shaft that reaches the upper level. Most of the ore mined consisted of oxidized material from the upper level which, according to M. J.

Lawler[10], generally averaged less than $16 per ton in gold.

As seen in the lower adit drifts, which are about 300 feet in length and 170 feet below the collar of the new shaft, the vein ranges from a few inches to about 1½ feet in width. It consists of stringers and irregular bunches of coarse-grained, massive white quartz with irregular bunches and disseminations of pyrite, galena, and sphalerite.

COWBOY MINE

The Cowboy property of four claims is accessible by one mile of road that branches westward from the Bagdad highway at a point 23 miles from Hillside. This deposit, which is reported to have been discovered in the eighties, was relocated in 1923 by G. G. Gray. The U. S. Mineral Resources credit it with a small production of gold-lead ore in 1911, 1925, and 1931. Some of this ore carried about an ounce of gold per ton.

The prevailing rock is micaceous schist, intruded on the east by granite and cut by granite-porphyry dikes. The principal vein strikes northwestward and dips about 60° SW. It has been opened by an inclined shaft, reported to be 200 feet deep, with about 700 feet of drifts. When visited in January, 1934, these workings were under water to the 65-foot level. So far as seen, the vein-filling consists of narrow, lenticular masses of brecciated jasper together with more or less coarsely crystalline shiny gray quartz. It contains small scattered masses and disseminations of limonite, cerussite, anglesite, and galena. The gold occurs mainly with the lead minerals, particularly in the jaspery portions of the vein.

CROSBY MINE

The Crosby property, in Secs. 4 and 9, T. 13 N., R. 8 W., is accessible from the Bagdad highway by 3.5 miles of road that branches westward at a point 13 miles from Hillside.

The U. S. Mineral Resources state that the Nieman and Crosby property produced in 1906-1907, 1911-1916, but do not give the amounts. In 1927, according to Carl G. Barth, Jr.[11], the Red Crown Mines, Inc., produced $1,000 in bullion, and lessees obtained $1,870 from 22 tons of ore. Some production was made during 1928 and 1930. In 1931, according to the U. S. Mineral Resources, 100 tons of ore that averaged more than 1.7 ounces of gold per ton were shipped, and 25 tons were treated by amalgamation and concentration. Lessees were continuing small scale operations in 1934. When visited in January, a little ore was being mined from the adit level and treated in an old 10-stamp amalgamation-concentration mill.

---

[10] Oral communication.
[11]Oral communication.

The mine is at an altitude of 3,300 feet, in a small area of banded gray schist that is surrounded by light-colored granite and intruded by pegmatite, rhyolite-porhyry, and basic dikes. The vein, which strikes N. 10° E. and dips 25°-30° E., ranges from less than an inch to about 18 inches in width. Its filling, where unoxidized, consists of coarse-textured, glassy, grayish-white quartz with bunches and disseminations of pyrite. Rich ore from the oxidized zone shows brecciated quartz with abundant cellular limonite. The gold appears to occur in the iron minerals and to a less extent as visible particles in the quartz. Considerable sericite has been formed in the wall rocks.

The vein has been opened by an incline, reported to be 350 feet deep, with water at the 235-foot level. According to Carl G. Barth, Jr., the vein has been largely stoped out for a length of 325 feet from the surface to the 165-foot level. He states that it is cut off on the south by a fault occupied by a basic dike.

SOUTHERN CROSS MINE

The Southern Cross Mine, in the southwestern part of the Eureka district, south of Grayback Mountain, is accessible by 4½ miles of road that branches southward from the Kingman road at a point 28 miles from Hillside. This deposit was opened by shallow workings more than thirty years ago. During the first few months of 1934, the present owner, R. L. Gray, shipped from the property about 55 tons of ore that is reported to have contained from 0.75 to 1.0 ounce of gold per ton.

The vein strikes northward, dips from 15° E. to almost flat, and occurs in vertical mica schist. Its gangue is coarse-textured, massive, grayish-white quartz with fractures and small cavities filled with limonite and sparse copper carbonate. The walls show marked sericitization and limonite staining.

Underground workings include a 70-foot inclined shaft that passes through the vein, and two short, near-surface drifts with small stopes on the vein. As shown by these workings, the vein is lenticular and ranges from a thin seam up to 4 feet in thickness.

MAMMOTH OR HUBBARD MINE

The Mammoth property of eight claims, held by Hugh Hubbard and associates, is 8½ miles by road north of Hillside and within ¾ mile of the Santa Maria River.

This deposit, which is reported to be on school land, was discovered many years ago. Since 1931, it has produced several cars of ore.

Here, a moderately hilly pediment is floored by extensively jointed granite. The present shaft, which is 175 feet deep, was sunk on a narrow southward-dipping fault zone that showed a little iron oxide and copper stain. Between the 80- and 100-foot levels, a short drift to the east encountered an ore shoot that

strikes S. 70° W., dips 40° NW., and is about 20 feet long by 2 to
2½ feet thick. This vein material consists of coarse-grained milky
quartz, pale-yellowish calcite, and fine-grained purple fluorite.
Small masses and disseminations of yellowish pyrite are present
in the quartz. In places, the pyrite is oxidized to limonite. The
wall rock shows strong sericitization. According to Mr. Hubbard,
the ore mined from this shoot in January, 1934, averaged about
0.4 per cent of copper, 0.51 ounces of gold, and 2 to 3 ounces of
silver per ton. Trucking to Hillside cost $1.50 per ton.

PRESCOTT DISTRICT

BULLWHACKER MINE

The Bullwhacker mine is about 4 miles in air-line east of Prescott and a short distance south of the Dewey road, on the divide
between Granite and Agua Fria creeks. The principal rocks are
dense black schists with dikes of diorite porphyry, intruded on the
west by Bradshaw granite. Blake[12], in 1898, described this deposit as "A small mine . . . sometimes called the Bowlder claim.
It is notable for bearing coarse gold of high grade in a small
quartz vein. The vein varies in thickness from a few inches to a
foot. The quartz is hard and occurs in bowlder-like masses,
rounded hard lumps, in which the gold occurs. There is apparently one ore chute or chimney pitching northward. The claim
has been worked to a depth of 132 feet by a shaft and most of the
pay ore extracted (1886) to that depth."
Lindgren[13] states that the massive milky-white quartz contains a little pyrite in crystals and stringers.

CHERRY CREEK DISTRICT[14]

The Cherry Creek district is in the southern portion of the
Black Hills, in the vicinity of Cherry post office, on the head-
waters of Cherry Creek. By highway, this place is 16 miles
from the railway at Dewey and 22 miles from Clemenceau.
Regarding the history and production, Lindgren says:
"Many of the mines, the Monarch property in particular, were
operated in a small way in the early days, their ore generally
being reduced in arrastres . . . In 1907 seven properties were in
operation, with six mills. Some high-grade ore containing as
much as $60 or even $100 to the ton was extracted. In 1908 six
mines yielded 464 tons, from which was obtained $5,775 in gold
and 86 ounces in silver, a total value of $12 to the ton. In 1909

[12] Blake, Wm. P., In Rep't. of Gov. of Arizona, 1898, p. 262.
[13] Work cited, p. 108.
[14] Lindgren, W., Ore deposits of the Jerome and Bradshaw Mountains
    quadrangles, Arizona: U. S. Geol. Survey Bull. 782, pp. 102-107, 1926.
[14] Reid, J. A., A sketch of the geology and ore deposits of the Cherry
    Creek district, Arizona: Econ. Geol., vol. 1, pp. 417-36, 1906.

four mines produced 330 tons yielding 329 ounces of gold and 127 ounces of silver, together with 29 tons of concentrates yielding 40 ounces of gold and 115 ounces of silver. In 1910 seven mines produced 1,332 tons, from which was obtained $6,352 in gold and 93 ounces of silver; this ore was obviously of low grade. In 1911 the district yielded $9,402 from 531 tons of ore, or about $17 to the ton. The producers were the Etta, Federal, Hillside, and Leghorn mines. In 1912 the Monarch and two other properties produced gold. In 1914 the production was $2,866 from four properties. In 1915 ore was mined from the St. Patrick, Garford, and Esmeralda claims. In 1916 two properties produced a little bullion ... In 1922 operations were again begun at the Monarch and the Logan." A little gold bullion was produced in the district during 1923 and 1925. Several cars of ore were shipped in 1930, 1931, 1932, and 1933.

Most of the district is in the upland basin of Cherry Creek, with elevations of 5,000 to 5,500 feet, but part of it extends down the steep eastern slope of the Black Hills. The prevailing rock is Bradshaw granite, locally overlain by Cambrian and Devonian sedimentary rocks and Tertiary lavas.

The veins occur in the granite, within shear zones which strike north-northeastward and dip at low or moderate angles westward. Their filling consists of irregular, lenticular bodies of massive, shiny white quartz with small amounts of greenish-black tourmaline. The ore is marked by irregular grains and bunches of more or less oxidized chalcopyrite, bornite, sphalerite, and galena. In places, pseudomorphs of limonite after pyrite are abundant. Although the water level is about 60 feet below the surface, oxidation, which is probably of pre-Cambrian age, extends to depths of 300 feet. The ore bodies are generally small. Part of the gold occurs as visible but fine particles in the quartz, particularly with limonite, but part is contained in the sulphides. Lindgren [15] states that the concentrates after amalgamation are reported to contain from 4 to 5 ounces of gold and a small amount of silver per ton. He regards these veins as positive examples of pre-Cambrian high temperature deposits. The Cherry Creek veins have yielded no placers of economic importance.

MONARCH AND NEARBY MINES[16]

The Monarch or Mocking Bird mine is at the eastern foot of the Black Hills, at an altitude of about 4,500 feet. It has been operated intermittently with stamp mills since 1886 and has probably produced more than any other mine in the district, but many of the old workings are caved. The country rock is fine-grained light colored granite which shows practically no alteration in the vein walls. The mineral deposit consists of several veins which

[15] Work cited, p. 103.
[16] Description abstracted from Lindgren, work cited, p. 105.

strike N. 10°-20°W. and dip 32°-45°W. They are made up of lenses, several feet in maximum width, of coarsely crystalline white quartz vein 5 to 6 feet wide, developed to a depth of 200 feet, ore is mostly free milling, but some galena and chalcopyrite are present.

The Etta, Gold Ring, and Conger mines, south of the Monarch, were producers during the eighties. The Conger is reported to have been recently worked in a small way by lessees. Lindgren says: "The Etta is mentioned in the Mint report for 1887 as a quartz vein 5 to 6 feet wide, developed to a depth of 200 feet, and containing ore of a value of $29 to the ton."

The Pfau mine, according to J. S. Sessions, about 2 miles south-southeast of the Monarch, produced intermittently for about nine years prior to 1904.

BUNKER OR WHEATLEY PROPERTY

The Bunker or Wheatly property of eight claims is a short distance northwest of the Inspiration ground and about 1½ miles north of Cherry. This property was worked to some extent in the early days. In 1923, it produced a little ore that was treated in the Federal mill. During 1932 and 1933, the present owner, E V. Bunker, shipped several cars of ore containing from 0.75 to 2.0 ounces of gold per ton. The principal workings are at an elevation of about 5,700 feet on three veins which dip gently southwestward and are from 25 to 45 feet apart. As exposed by the present shallow workings, these veins range up to 6 feet, but probably average less than one foot, in thickness. Considerable massive quartz is present. The gold occurs, very finely divided and associated with abundant limonite, within cellular and brecciated quartz.

GOLDEN IDOL OR HILLSIDE MINE

The Golden Idol or Hillside mine is 1½ miles by road north of Cherry, at an altitude of about 5,400 feet. Lindgren[17] states that the property was worked from 1907 to 1910 and was equipped with a stamp mill and cyanide plant. During the past fifteen years, it has been held by the Verde Inspiration Company and the Western States Gold Mining Company, but has made little or no production. Lindgren continues: "There appear to be three veins on the property, and on one of them an incline 375 feet long has been sunk at a dip of 35° W. . . . Pits near the shaft show a 4-foot vein of sheared granite with bunches of quartz. The quartz shows bluish-black streaks of tourmaline, also a little pyrite and chalcopyrite. It contains solution cavities with limonite. The ore is said to have contained $7 to $12 to the ton."

---

[17] Work cited, p. 106

FEDERAL MINE

The Federal mine is west of the Bunker, about 1¼ miles north of Cherry at an altitude of 5,300 feet. Its southward-dipping vein is reported to have been explored by a 260-foot incline in 1907. A mill was built at about that time, but little ore was mined.

LEGHORN MINE

The Leghorn mine, about 1¾ miles north of Cherry, is reported to have been worked intermittently, with some production, from 1904 to 1918, and to a small extent in 1924[18]. Lindgren[19] says: "The vein is contained in granite and has been opened by an incline 600 feet long, dipping 35° W. In Weed's Mines Handbook for 1922 it is stated that there are 6,000 feet of workings. A Chilean mill has been erected on the property. . . . The vein is said to average 2 feet in width. The quartz contains chalcopyrite and gold, but it is probable that difficulties were encountered below the zone of oxidation. Specimens from the dump show abundant solution cavities filled with hematite and secondary quartz."

GOLD BULLION OR COPPER BULLION MINE

The Gold Bullion, formerly know as the Copper Bullion property, is about 2 miles west-northwest of Cherry. During the early days, according to local reports, it was opened by a 660-foot incline and several hundred feet of shallower workings. These openings were on a steeply westward-dipping vein that pinches and swells to a maximum width of about 7 feet. As seen near the surface, it consists of lenses of quartz together with locally abundant masses of hematite and limonite. The gold is very finely divided. In places, the quartz contains irregular bunches of partially oxidized galena. Copper stain is locally present. Since 1930, several cars of shipping ore have been mined from near the surface.

GOLD COIN MINE

The Gold Coin property, which in 1934 was being worked by the Southwestern Gold Mining Corporation, is east of Hackberry Wash, about ¼ mile from the Dewey road. In the early days, this property was opened by a shaft about 100 feet deep. Within the past two years, it has been developed by a 118-foot shaft and has produced several cars of ore. The vein dips steeply eastward, is rather pockety, and attains a maximum width of about 3 feet.

QUAIL AND GOLDEN EAGLE MINES

Some ore has recently been shipped from shallow workings on lenticular, steeply eastward-dipping veins on the Quail and Golden Eagle groups which are adjacent to the Dewey road and Hackberry Wash.

[18] Oral communication from J. S. Sessions.
[19] Work cited, p. 107.

ARIZONA COMSTOCK OR RADIO MINE

The Arizona Comstock or Radio group, east of Hackberry Wash, is reported to have produced some ore from near the surface during the early days. It shows a steeply southwestward-dipping vein, up to about 20 inches wide, that was opened by a shallow shaft, a winze, and about 100 feet of drifts.

GOLDEN CROWN MINE

The Golden Crown property is east of Hackberry Wash and southeast of the Dewey road. Early in 1934, it was being worked by Binder Brothers, and is reported to have shipped two cars of ore during 1933. The vein, which is rather irregular, dips approximately 25° SW. and ranges up to about 3½ feet in thickness. It has been opened by about 300 feet of drifting from a shallow incline that encountered considerable water at 50 feet. Its quartz is massive to brecciated and contains abundant limonite derived from coarse-grained pyrite.

LOGAN MINE[20]

The Logan mine, about 2 miles southwest of Cherry, was reopened in 1922 and operated for a short while by the New United Verde Copper Company. It was idle in 1934. Material on the dump consists of decomposed granite and slightly copper-stained quartz. The property was equipped with a small mill.

GROOM CREEK DISTRICT

The Groom Creek district is mainly in the vicinity of upper Groom Creek, an intermittent stream that flows southwestward to join the Hassayampa at a point about 5 miles south of Prescott. Within its drainage area, which ranges in elevation from 5,400 to more than 7,000 feet above sea level, are several silver-gold-bearing quartz veins which have been worked intermittently for many years. In this area, water and timber are relatively abundant, but operations during winter are sometimes hampered by snow.

The principal rocks are pre-Cambrian sedimentary schist, intruded by stocks and dikes of granodiorite and diorite. The quartz veins tend to be narrow and lenticular. They probably belong to the mesothermal type but have been worked mainly above the sulphide zone.

The Midnight Test (National Gold), Empire, King-Kelly-Monte Cristo, Victor, and Home Run properties are in this district. When visited in January, 1934, only the Midnight Test mine was being actively operated.

NATIONAL GOLD (MIDNIGHT TEST) MINE

The Midnight Test mine, held by the National Gold Corporation, is on the northwestern slope of Spruce Mountain, at an

---

[20] Abstracted from Lindgren, W., work cited, p. 107.

elevation of 7,000 feet. According to local people, this mine was worked to shallow depths during the early days and was developed by a 400-foot shaft prior to 1906. For a short while during this period, a small mill operated on the property. In 1919, the mine yielded some milling ore. The total production prior to 1922 is reported to be $100,000[21], part of which came from a rich shoot of silver ore near the surface. Subsequently, the property was obtained by the National Exploration Company, later the National Gold Corporation, which has carried on considerable underground exploration and built a 200-ton mill designed for amalgamation, flotation, and table concentration. The National Gold Corporation has made several shipments of ore and concentrates.

The principal rocks of the vicinity consist of silicified schist, intruded by stocks and dikes of granodiorite and diorite. When visited in January, 1934, the main shaft was 600 feet deep, on an incline of 80° W., and the workings seen followed a shear zone that trends N. 15°-20° W., mainly in the schist. The fractured portions of this zone are generally marked by abundant limonite and hematite. No manganese dioxide was seen. In some places, irregular to lenticular, generally narrow veins of coarse-grained, druzy quartz are present. This quartz contains scattered kernels of galena and, mainly below the 300-foot level, bands and scattered bunches of pyrite with minor amounts of sphalerite. Gold is reported[22] to occur mainly in the iron oxides of the shear zone and to a less extent in the quartz and sulphides.

OTHER PROPERTIES

The Monte Cristo mine, one-half mile east of Groom Creek settlement, is reported to have produced considerable silver ore and some gold during the eighties, from 1902 to 1905, and in 1920.

Lindgren states that the King-Kelly fissure, which strikes N. 15° W., with vertical or steep westerly dip, is a narrow vein containing fine-grained druzy quartz, sparsely disseminated pyrite, and a little galena.[23]

The Empire vein, according to Lindgren, outcrops in a quartzitic schist and diorite and strikes N. 20° W. He states that, from about 1902 to 1910, a considerable amount of ore containing approximately equal values of gold and silver was mined from the oxidized zone and milled. A 300-foot shaft, sunk to water level, showed that the ore becomes pyritic in depth.[24]

The Home Run property is reported to have shipped a car of ore in 1932.

---

[21] Lindgren, Waldemar, Ore deposits of the Jerome and Bradshaw Mountains quadrangles, Arizona: U. S. Geol. Survey Bull. 782, p. 114.

[22] Oral communication from W. W. Linesba.

[23] Work cited, p. 113.

[24] Work cited, p. 114.

WALKER DISTRICT[25]

The Walker district is near the head of Lynx Creek, in a well-watered, wooded region more than 6,000 feet above sea level. Here, many veins with free gold in the oxidized zone were discovered and worked with arrastres by the early-day placer miners. Lindgren estimates that the total production of the district does not exceed $1,500,000. More than half of this yield was made prior to 1900.

The productive veins are mainly in a granodiorite stock, 2 miles long by one mile wide, that intrudes Yavapai schist on the northwest and Bradshaw granite on the southeast. All of these rocks are cut by prominent dikes of rhyolite-porphyry. The veins, which occur within steeply dipping fault zones several feet wide, commonly consist of several streaks of quartz with gold-bearing sulphides.

SHELDON MINE[26]

The Sheldon mine, about a mile southwest of Walker, is estimated to have produced about $200,000 prior to 1922. The vein, which is in the southwestern portion of the granodiorite area, strikes N. 30° E., dips from 70° to 80° SE., and is traceable for more than half a mile on the surface. Prior to 1922, it had been opened by a 650-foot vertical shaft with several hundred feet of drifts on the 250, 450, and 650-foot levels, and oxidized ore had been stoped in places between the 200-foot level and the surface. In 1922, the mine made about 185 gallons of water per hour. A 200-ton concentration flotation plant was built in 1924 and several thousand tons of copper-lead ore, carrying some gold and silver, was treated in 1925, 1926, 1929, and 1930. During this period, the shaft was deepened to 1,280 feet and several thousand feet of development work was done.

The vein pinches and swells but averages 4 or 5 feet in thickness. Its principal ore shoot apparently pitches 60° northward and is reported to be 16 feet wide by 700 feet long on the 650-foot level. The vein minerals are milky-white, vuggy quartz and some calcite, with more or less pyrite, sphalerite, chalcopyrite, galena, and tetrahedrite. Some supergene copper minerals are present in the oxidized zone. The quartzose ore is stated to contain an ounce or more of gold per ton, but the sericitic and pyritic granite is only about one-tenth as rich. The ore mined in 1923 was reported to average about 2.76 per cent of copper, 3.5 per cent of lead, $5 in gold, and 8.5 ounces of silver per ton. In 1931, the output amounted to about $18,000 in gold, 78,400 ounces of silver, 908,377 pounds of copper, and 36,693 pounds of lead, in all worth about $170,000.

---

[25] Abstracted from Lindgren, W., work cited, pp. 109-10.
[26] Abstracted largely from Lindgren, W., work cited, pp. 110-12.

MUDHOLE MINE[27]

The Mudhole mine, a short distance southwest of Walker, was worked to some extent prior to 1897. Its production from 1897 to 1903 is reported to have been $480,000 in gold and silver. The property has been practically idle since 1912.

This deposit, which appears to be mainly in dark magnetitic hornfels, near the contact of the granodiorite stock, consists of two parallel veins, each 6 to 8 feet wide. Workings include a 740-foot shaft, inclined at 47°, and a shaft and tunnel about 2,000 feet farther southwest. Bleached hornfels on the dump at this tunnel shows seams which contain galena and sphalerite with some chalcopyrite and pyrite. The ore is reported to have contained from $7 to $15 in gold and silver.

BIGBUG DISTRICT[28]

The Bigbug district is on the northeastern slopes of the Bradshaw Mountains. It ranges in altitude from 7,000 feet, west of Bigbug Mesa, to 4,500 feet, in Agua Fria Valley. The western portion is timbered and fairly well watered, while the lower dissected pediment or foothill belt is rather dry and brushy to open country.

This area is made up of schist, intruded in places by diorite, granodiorite, granite. and dikes of rhyolite-porphyry. The schist is mainly of sedimentary origin, with many quartzitic beds, but contains also some igneous members. It is intruded on the west by the Mount Union belt of granite, and southwest of McCabe, by a stock of granodiorite. These relations are shown on the geologic map of the Bradshaw Mountains quadrangle, by T. A. Jaggar and C. Palache.[29] Basalt flows of post-mineral age form Bigbug Mesa where they rest upon a late Tertiary or early Quaternary pediment. Elsewhere in the district, this pediment has been extensively dissected by post-basalt erosion.

Lindgren has classified the ore deposits, other than placers, as follows: (1) Pyritic copper deposits, such as the Blue Bell, Hackberry, Butternut, and Boggs; (2) Pre-Cambrian quartz veins, such as the old Mesa, near Poland; (3) The Iron King gold-silver replacement deposit; (4) Later veins, probably connected genetically with rhyolite-porphyry dikes, mainly near Poland and Providence.

During the early days, some of the Bigbug deposits yielded a considerable amount of gold and silver from the oxidized zone. From 1901 to 1931, inclusive, the production of the district, as recorded by the U. S. Mineral Resources, amounts to approximately $17,000,000 in copper, gold, silver, lead, and zinc. Nearly $4,000,000 of this amount was in gold of which about $30,000 came from placers.

---

[27] Abstracted from Lindgren, W., work cited, p. 112.
[28] Largely abstracted from Lindgren, W., U. S. Geol. Survey Bull. 782.
[29] Published by U. S. Geol Survey in Folio 126 and Bulletin 782.

IRON KING MINE[30]

"A little more than a mile west of the Humboldt smelter, in the open foothills, is the Iron King mine, now owned by the Southwest Metals Company, which also owns the Humboldt smelter. To the officers of that company I am indebted for most of the following information. The deposit, which carries gold and silver, forms a replacement zone in the Yavapai schist, but it differs from the normal copper deposits that are so numerous farther to the south in the same schist. It was worked about 1906 and 1907. The production in 1907 was 1,253 ounces of gold, 35,491 ounces of silver, and 3,933 pounds of copper.

"The deposit is developed by two shafts 750 feet apart and 435 and 225 feet deep. Several thousand tons of ore averaging $8 a ton in gold and silver have been shipped to the neighboring smelter. It is claimed that the ore in sight amounts to 20,000 tons and that the deposit contains much low-grade siliceous material averaging $1 or $2 in gold to the ton. The ore is reported to contain from $6 to $8 in gold and 4 to 23 ounces in silver to the ton. Some diamond drilling has been done; the cores in the ore body contained $8 in gold and 9.60 ounces of silver to the ton, 32 per cent of iron and 14 per cent of insoluble matter. Other parts of the ore body contain as much as 70 per cent of insoluble constituents.

"The deposit forms a series of lenses in part overlapping, in highly silicified schist, which strikes N. 21° E. and dips 75° W. These lenses are 150 to 500 feet long and 5 to 10 feet wide. The whole mineralized zone is 75 feet wide.

"The water level was found at a depth of 140 feet, and near this level in one ore body there was some enriched copper ore containing 4 to 5 per cent of copper.

"The ore is a steel-gray flinty schist containing a crushed quartz mosaic of coarser and finer grain intergrown with some dolomitic carbonate and abundant prisms of bluish-gray tourmaline. The sulphides are disposed in streaks and consist of fine-grained arsenopyrite, pyrite, light-colored sphalerite, and a little chalcopyrite and galena."

McCABE-GLADSTONE MINE[31]

The McCabe-Gladstone property of eight claims is a short distance south of McCabe, on Galena Gulch.

During the early seventies, this deposit yielded considerable amounts of rich oxidized ore. The property then remained practically idle for many years. It was worked continuously from 1898 to 1913 by the Ideal Leasing Company, with a reported production of $2,500,000 to $3,000,000. The mine was again idle from 1913-1933 but early in 1934 was reopened and unwatered by H.

---

[30] Quoted from Lindgren, W., work cited, pp. 127-28.
[31] Largely abstracted from Lindgren, W., work cited, pp. 130-32.

Fields and associates. In June, 1934, mine ore, mixed with old gob and dump material, was being treated in a 200-ton flotation mill.

The mine is developed by two shafts, 800 feet apart and 900 to 1,100 feet deep, together with several miles of workings. A longitudinal section of the mine is given in U. S. Geol. Survey Bulletin 782.

Here, amphibolitic schist is intruded by dikes of rhyolite-porphyry and, a short distance farther southwest, by a stock of quartz diorite.

The vein strikes N. 54° E. and dips 79° SE., but, between the two shafts, a 20-foot dike of rhyolite-porphyry apparently deflects the strike southward. The vein averages about 3½ feet wide. Stoping has followed five ore shoots, each 200 to 500 feet long. At least two of them appear to extend to the 1,100-foot level. They pitch steeply westward and average somewhat less than a foot in thickness.

The ore consists of quartz together with considerable amounts of pyrite and arsenopyrite and a little sphalerite, galena, and chalcopyrite. The following analysis of the shipping ore and the concentrates is given: Silica, 31.4 per cent; copper, 2.0 per cent; lead, 2.1 per cent; zinc, 4.7 per cent; iron, 24.6 per cent; arsenic, 3.9 per cent; antimony, 1.0 per cent; sulphur, 20.4 per cent; gold, 1.6 ounces per ton; silver, 10.2 ounces per ton.

UNION MINE[32]

The Union mine is about 1¾ miles southwest of McCabe, in the upper part of Chaparral Gulch, at an elevation of approximately 5,000 feet.

This deposit, which became known in the late sixties, at one time was consolidated with the Little Jessie. Except for a little intermittent work and small production, the property has been practically idle for many years. Early in 1934, the Union and Jessie mines were reported to be held by the Arizona Consolidated Mining Company which was carrying on development work and installing new milling machinery.

The workings include a 1,200-foot tunnel, with more than 1,000 feet of drifts on the vein, and a 200-foot shaft sunk from the tunnel level. The vein, which is a continuation of the Lelan vein, strikes about N. 70° E., dips steeply southeastward, and is followed by a later unmineralized basic dike. The ore consists of massive glassy quartz, up to 10 feet thick, with irregularly disseminated pyrite, arsenopyrite, sphalerite, and galena. Where cut on the tunnel level and on the 77-foot level of the shaft, the ore shoot is reported to be 250 feet long, with a pitch of about 30° SW. The lower limit of the ore is reported to be about half an ounce in gold per ton. Except in the oxidized zone, which is shallow, the gold does not occur free.

---

[32] Largely abstracted from Lindgren, W., work cited, pp. 133-34.

LITTLE JESSIE MINE

The Little Jessie mine is about 1,700 feet south of the Union. This deposit was discovered in 1867. From about 1890 to the end of 1898, it was worked by J. S. Jones and lessees. Their mill is reported to have produced about $750,000 worth of bullion and concentrates, chiefly from the Little Jessie. From about 1909 to 1916, considerable development work was done and a little ore was shipped, mainly by the Chaparral Mining Company. Early in 1934, the Arizona Consolidated Mining Company was reported to be carrying on development work and installing new mill machinery at the Union-Jessie property.[33]

Lindgren states that, in 1922, the shaft was 659 feet deep, and that much high-grade auriferous pyrite was encountered between the 500- and 600-foot levels. He adds that the ore contains from one-half to one ounce of gold per ton and very little silver.[34]

LELAN-DIVIDEND PROPERTY

The Lelan mine is on a ridge southwest of the Jessie.

This deposit was discovered during the sixties. Browne's report for 1868 states that 60 tons of ore from the Dividend mine, treated in the Big Bug (Henrietta) mill, yielded $20 per ton in free gold.[35] At that time however, it was not of commercial grade. According to Lindgren, the Lelan and Dividend were worked more or less from 1900 to 1914, and during part of that time were equipped with a 10-stamp mill. He states that their ore production prior to 1923 was probably at least 10,000 tons which contained from a half to 3 ounces of gold per ton, together with a little silver, copper, and lead.[36] In 1932 and 1933, the property was operated by the Southern Exploration Company with a force of about twenty-five men. This company erected a 100-ton flotation-concentration plant and produced concentrates during part of 1933. Operations were suspended at the end of the year.

The vein, which is a continuation of the Union, strikes northeastward and dips steeply southeastward. It is opened by a 500-foot shaft inclined at 80°, with development on five levels. Most of the recent production is reported to have come from the fourth level. The vein is rather lenticular and ranges up to several feet in width. Its filling consists of massive, shiny white quartz with irregular masses, seams, and disseminations of pyrite, chalcopyrite, sphalerite, and galena. The gold occurs in the sulphides.

---

[33] History compiled by J. B. Tenney.
[34] Work cited, pp. 132-33.
[35] Brown, J. Ross, Mineral resources of the states and territories west of the Rocky Mountains. 1868.
[36] Work cited, p. 133.

HENRIETTA OR BIG BUG MINE

The Henrietta mine which, in the early days, was known as the Big Bug, is about one-half mile north of Bigbug Creek and one mile west of Poland siding.

Browne's report for 1868 states that, in 1866, the Big Bug mine was some 50 feet deep and was producing ore from near the surface. In 1871, according to Raymond, the Big Bug vein was not being worked, but the Big Bug 10-stamp mill was treating gold ores from the vicinity.[37] At that time, the combined costs of mining and milling amounted to about $9 per ton.

The following data are largely abstracted from Lindgren's report: In 1883 and 1884, the Big Bug property was the most prominent one in the district. During this early period, the mine made a large production mainly from the oxidized gold ores from the upper levels. From 1915 to 1919, the mine was operated by the Big Ledge Copper Company which did considerable development below the old workings and produced gold-bearing copper ore. The property was equipped with a 100-ton flotation mill. In 1923, this company was reorganized as the Huron Copper Mining Company. Some shipments of copper ore containing gold were made in 1926 and 1930.

A longitudinal section of the workings is shown in U. S. Geol. Survey Bulletin 782. The old developments, which extended to the sulphide zone, include a 500-foot shaft, on the ridge, with a 1,500-foot tunnel through the ridge, 220 feet below the collar, and considerable stoping. Farther north, on the Gopher claim, the vein has been opened to depths of a few hundred feet. The deeper work, which was done by the Big Ledge Copper Company, included a 2,200-foot tunnel and a 600-foot winze with levels and stopes extending a few hundred feet northward.

The vein, which occurs mainly in massive, fine-grained amphibolite or diorite, strikes north, dips about 70° W., and is from 2 to 6 feet wide. Its gangue consists of massive quartz with some calcite. About 60 per cent of the unoxidized ore consists of pyrite, chalcopyrite, sphalerite, and galena. Ore from the lower levels is reported to contain 3.2 per cent of copper and 14 per cent of iron, together with 0.2 ounces of gold and 2.7 ounces of silver per ton.

POLAND-WALKER TUNNEL[38]

Poland, at the northern foot of Bigbug Mesa, is accessible by road from the Black Canyon Highway. The spur of the Santa Fe Railway that formerly served this vicinity was dismantled a few years ago. Near the southern portal, amphibolite is intruded on the north by somewhat schistose granite, and on the west by a 75-foot dike of rhyolite-porphyry. The tunnel extends northward

---

[37] Raymond, R. W., Statistics of mines and mining in the states and territories west of the Rocky Mountains. 1871.

[38] Largely abstracted from Lindgren, W., work cited, p. 136.

for 1,100 feet through a ridge of this granite. It exposed several veins upon which considerable work has been done.

The Poland vein, which was cut 800 feet from the south portal, strikes northeast and dips steeply northwest. Ore on the dump shows druzy quartz with pyrite, sphalerite, and galena. According to local reports, the vein was opened by several thousand feet of drifts and a 325-foot shaft below the tunnel level. From 1900 until about 1912, intermittent production was made with a 20-stamp mill. The 1907 yield was $130,465 in gold and 16,609 ounces of silver. The total output for this period is estimated at $750,000, probably mostly in silver. According to the U. S. Mineral Resources, the mine made a small production of gold ore in 1926, 1930, and 1931. Early in 1934, occasional shipments of gold-bearing ore and concentrates were being made by F. Gibbs and associates.

Prior to 1922, some production was made from the Occidental vein which is reported to have been cut 500 feet from the north portal of the tunnel and followed to a depth of 200 feet below the tunnel level. This vein, which is said to be similar to the Poland vein, carries gold, silver, and lead.

MONEY METALS MINE

The Money Metals mine, about 1¼ miles west of the Poland tunnel and Bigbug Mesa, is accessible by a road that branches northeastward from the Senator Highway at a point about ⅛ mile south of the Hassayampa bridge.

This deposit was located in 1897 by F. Reif who shipped some ore from the upper levels and sold the property. After some further development work, the mine remained idle until 1928 when it was reopened. Since 1933, it has been operated by the Yavapai Gold and Silver Mining Company.

The country rock is gneissoid granite. A rhyolite-porphyry dike about 60 feet wide follows the hanging wall of the vein, and, a short distance farther west, a mass of diorite intrudes the granite.

Workings on the property include a 300-foot shaft, inclined 68° W., together with a total of approximately 1,400 feet of drifts on three levels. When visited in February, 1934, water was kept from the 200-foot level with a Cornish pump.

As exposed underground, the vein strikes N. 50° W. and dips 65° to 70° W. In places, it has been offset by transverse faults. The vein filling consists of coarse-grained, grayish white quartz with irregular masses, veinlets, and disseminations of galena, sphalerite, pyrite, and chalcopyrite. The wall rock shows strong sericitic alteration. According to J. K. Kilfeder,[39] mine superintendent, much of the vein contains about half an ounce of gold per ton. On the 200-foot level, the ore shoot is about 175 feet long by 2 to 5 feet wide.

---
[39] Oral communication.

Surface equipment on the property includes a 20-ton concentrator powered with two Dodge motors. The sulphide concentrates are reported to carry more than $200 in gold per ton.

HASSAYAMPA DISTRICT[40]

The Hassayampa district, on the western slopes of the Bradshaw Mountains, is southwest of the Walker and Groom Creek districts and northwest of the Turkey Creek district.

These slopes, which have been carved into narrow, steep-sided canyons by the tributaries of Hassayampa Creek, descend from an elevation of 7,900 feet to 4,500 feet. The upper portions are well forested, and the lower reaches support dense brush. Except during dry seasons, the principal canyons carry water. During winter, the higher country is subject to heavy snows which may block the roads for weeks at a time.

Raymond, in 1871, stated[41] that the district was first visited and organized by prospectors in the spring of 1864, originally to work the placers only, but subsequently a large number of quartz veins were discovered and located.

During the early days, these veins were important producers of gold and silver from the rich, though shallow, oxidized zone. Since 1895, the sulphide zone has yielded considerable gold, silver, copper, lead, and zinc. From 1904 to 1931 inclusive, the U. S. Mineral Resources record a total production of $1,104,491 of which $469,940 was in gold. Part of this total came from the Groom Creek district, and a small portion of the gold was from placers. During the past few years, several of the old mines with gold-bearing shoots have been reopened.

These deposits occur mainly in a belt of schist that extends northeastward between the granite areas of Granite Creek and Mount Union, although some are in the granite. The schist has been intruded by diorite, and the granite and schist by dikes of rhyolite-porphyry.

According to Lindgren's classification, some of the veins, as the Ruth and Jersey Lily, are believed to be of pre-Cambrian age, while others, as the Senator and Venezia deposits, are associated with rhyolite-porphyry dikes and appear to be of Mesozoic or Tertiary age.

ORO FLAME AND STERLING MINES

Situation and history: The Oro Flame-Sterling group of seven patented and twelve unpatented claims is in the vicinity of Hassayampa Creek, about 6 miles south of Prescott. The camp is accessible by 2 miles of road that branches south from the Wolf Creek road at a point 2 1/3 miles from U. S. Highway 89.

---

[40] In part abstracted from Lindgren, W., U. S. Geol. Survey Bull. 782, pp. 114-16. 1926.

[41] Raymond, R. W., Statistics of mines and mining in the states and territories west of the Rocky Mountains, 1871, pp. 238-41.

In 1871, Raymond said that "The Sterling mine has become quite famous, as much on account of the richness of the sulphurets it contains as from the repeated failures in working them. It was discovered in 1866."[42] He stated that an unsuccessful attempt had been made to treat the ore with a 10-stamp mill equipped with amalgamation plates, Hungerford concentrators, and chlorination apparatus.

Prior to 1908, according to H. K. Grove, the Oro Flame, then known as the Mescal claim, made a considerable production with a 20-stamp mill.[43]

Since 1928, the Oro Flame-Sterling property has been worked by H. K. Grove and associates, the Oro Flame Mining Company and the Oro Grande Mining Company. The U. S. Mineral Resources states that about 1,000 tons, shipped in 1929, contained 913.32 ounces of gold, 2,529 ounces of silver, and 12,676 pounds of copper. More than 800 tons were shipped in 1930. According to Mr. Grove, the total output from 1928 to 1933 amounted to eighty cars of ore that averaged $25 per ton. Most of this ore came from the Oro Flame workings. A 40-ton flotation and concentration mill was completed early in 1934.

Topography and geology: In this vicinity, the deep, meandering canyon system of Hassayampa Creek exposes schist, intruded by large masses of diorite and granite and persistent dikes of rhyolite-porphyry.

Veins: In the southern portion of the property, the Oro Flame vein occurs within a fault zone that strikes N. 20° W., dips 76° NE., and separates gneissic granite on the northeast from schist with diorite on the southwest. On the southeastern side of Hassayampa Creek, the vein has been opened by a 320-foot inclined shaft, an adit, and several hundred feet of drifts. Above the 220-foot level, it has been largely stoped out for a length of 400 feet by a width of 3½ feet. The mine makes but little water. As shown by these workings, the fault zone is from 3 to 6 feet wide, and the ore occurs mainly as narrow vertical lenses, seams, and bunches that trend northward, more or less diagonally from foot wall to hanging wall. The best ore shoots seem to occur where rather flat fractures intersect the footwall. The ore consists of massive grayish-white quartz with irregular masses, veinlets, and disseminations of fine-grained galena and fine-grained, pale-yellowish pyrite. The gold occurs mainly in the sulphides, particularly the galena. According to Mr. Grove, the ore carries about 0.12 per cent of copper and 3 ounces of silver to each ounce of gold. In the semi-oxidized zone, which in the shaft extends to a depth of about 60 feet, some free gold, accompanied by hematite and limonite, was present. The vein wall rocks show considerable alteration to sericite and carbonate.

---

[42] Raymond, R. W., Statistics of mines and mining in the states and territories west of the Rocky Mountains, pp. 240-41. Washington, 1872.
[43] Oral communication.

Near the new mill on the northwestern side of Hassayampa Creek, the vein has been opened by the 110-foot Oro Grande shaft. Some milling ore was stoped from near the surface.

About 150 feet east of the Oro Flame adit, a vein strikes about N. 20° W., dips 80° SW., and outcrops near a persistent dike of rhyolite-porphyry. Both this vein and the Oro Flame vein are reported to be traceable northward across the property. At the northern or Sterling end, their supposed continuation appears as two veins that strike northeastward and dip steeply towards each other. According to Mr. Grove, the western or Gold Bug vein has been opened by a 410-foot shaft, inclined at 45°, with more than 1,000 feet of drifts on the 300-foot level. Water now stands at the 250-foot level. Mr. Grove states that the eastern or American Eagle vein has been opened by two surface tunnels, a little stoping, and a drift from the 300-foot level of the Gold Bug workings. The Sterling ore is massive white quartz with practically no sulphides other than the pyrite and chalcopyrite. Raymond described the Sterling (Gold Bug) outcrop and workings of 1870 as follows: "It occurs in greenstone and metamorphic slates, parallel to which it strikes northeast, and dips with them to the southeast. There are very large croppings of brown-streaked quartz on the surface, which have yielded in the mill belonging to the company from $15 to $20 per ton. The vein is opened by an incline 118 feet deep. The largest body of ore was encountered from the surface to a depth of 53 feet, where the quartz was 16 feet wide and filled with iron and copper sulphurets, the former largely predominating. This chimney continued of the same size for 100 feet along the strike of the vein as far as explored, but in depth gave out below the 53-foot level." He states that a 100-ton lot of this ore, of which about 10 per cent was sulphides, yielded $15 per ton in free gold and $6 per ton by chlorination.[44]

RUTH MINE[45]

The Ruth deposit, on Indian Creek, ¾ mile north of Hassayampa Creek, is opened by a 300-foot shaft. It has yielded some gold, but its later production, which was made in 1911-1913, 1916, and 1926, has been chiefly lead, zinc, and silver ore and concentrates. This vein occurs in Bradshaw granite which, near the walls, is schistose, soft, probably sericitized, and impregnated with tourmaline and pyrite.

The vein dips steeply eastward. It consists of coarse-grained milky quartz, with narrow seams of pyrite, ankerite, and tourmaline. Pyrite, chalcopyrite, galena, and sphalerite occur as irregular bunches and streaks in the quartz.

---

[44] Work cited, p. 240.
[45] Abstracted from Lindgren, work cited, p. 116.

JERSEY LILY MINE[46]

The Jersey Lily mine is 4 miles south of the Ruth, on a ridge at an altitude of 6,000 feet. The production, which is said to amount to $7,000 worth of gold, was made many years ago.

In this vicinity, the rocks are slaty and fissile to amphibolitic schists. The vein, which is thick, consists of massive milky quartz with small crystals and thin, irregular veinlets of pyrite. Although rich in scattered spots, it is generally of low grade.

DAVIS-DUNKIRK MINE

The Davis-Dunkirk mines, held by Davis-Dunkirk Mines, Inc., is on the western slope of the Bradshaw Mountains, near the head of Slate Creek. By road, the main camp, at the mill, is 14 miles from Prescott and 3 miles west of the Senator Highway. As the elevation at the camp is about 6,400 feet and on part of the road over 7,000 feet, communications in winter are sometimes hampered by snow.

These veins were located in the sixties or early seventies of the past century. Raymond, in 1874, stated that: "Seven and a half tons of ore from the Davis mine near Prescott have been shipped to San Francisco and yielded $618.75 or $88.50 per ton. This vein is only opened by a few holes from 5 to 10 feet deep."[47] Little other historical data for the mine are available, but its production prior to 1922[48] is said to have been about $200,000. In 1925, the property was obtained by Davis-Dunkirk Mines, Inc., which subsequently built a new camp and a 120-ton flotation mill. Regular operations, employing about twenty men, began in August, 1933, and yielded fifteen cars of concentrates up to January 10, 1934.

Here, the steep-sided canyon of Slate Creek exposes moderately fissile pre-Cambrian schist, intruded by stocks and dikes of granodiorite and diorite and a few dikes of rhyolite-porphyry. The mine workings are mainly in schist which, a short distance farther west, is intruded by a large stock of granodiorite and, on the southeast, by diorite.

The Dunkirk vein occurs within a fault fissure that strikes northeastward and dips nearly vertically southeastward. As seen on the lowest adit level, its thickness ranges from a few inches to about 3 feet. The filling is mainly coarse-grained grayish-white quartz with abundant pyrite and chalcopyrite. In places, coarsely crystalline pale-pinkish ankerite is the principal gangue. Rather intense sericitization and silicification are apparent in the wall rock. The richer gold and silver ore shoots, as indicated by several old stopes above this adit, tend to be somewhat irregular

---

[46] Abstracted from Lindgren, W., work cited, p. 117.

[47] Raymond, R. W., Statistics of mines and mining in the states and territories west of the Rocky Mountains, p. 347. Washington, 1874.

[48] Lindgren, W., work cited, p. 119.

in form and distribution. The silver content is asserted to be decreasing with depth. At a point approximately 1,400 feet in from the portal, the vein separates into two branches about 45° apart. In January, 1934, ore was being mined from workings at the bottom of a 100-foot winze sunk at this point. According to H. L. Williams,[49] manager of the property, the mill concentrates from this ore averaged about one ounce in gold and 62 ounces in silver per ton.

The left branch of the vein continues northeastward, towards the old Davis mine. Ore from the upper levels of this mine was described by Lindgren[50] as consisting of fine-grained, druzy quartz with sparse pyrite and yellow sphalerite, with abundant grains of proustite and polybasite. He regarded this vein as of the shallow-seated type, but the Dunkirk vein belongs distinctly to the mesothermal type.

Workings on the Davis-Dunkirk property consist of more than 6,000 feet of adits, raises, stopes, and winzes, distributed over about four claim-lengths and with a vertical range of 1,600 feet.

TILLIE STARBUCK MINE

The Tillie Starbuck mine, held by H. L. Williams and associates, is near the head of Slate Creek Canyon, less than a mile east of the Davis-Dunkirk mill. This property was formerly owned by Major A. J. Pickrell who developed it with a few thousand feet of tunnels, winzes, and raises.

Lindgren[51] gives the following description: "The country rock is Yavapai schist intruded by dikes of light-colored rhyolite-porphyry. The foot-wall is said to be followed by a dike of rhyolite-porphyry 10 feet wide. The strike of the vein is N. 10° W., the dip 80° E., and the width 2 to 17 feet. The outcrop is persistent on the high ridge to the south, where the ore is largely oxidized. There are three ore shoots with backs of about 700 feet above the lowest tunnel level. This lowest tunnel is first a cross-cut southeast to the vein for 640 feet and continues on the vein for 1,000 feet. It is claimed that 100,000 tons of $10 ore have been developed in the vein.

"The ore, which is mainly quartzose, contains from $10 to $17 to the ton, of which two-fifths is gold and three-fifths silver.[52] The ore carries free gold and pans colors. The quartz is rather fine grained, is milky with many small druses, and includes numerous sericitized rock fragments ... The ore minerals are sparse pyrite and sphalerite in small grains and in places specks of pyrargyrite, which appears to be of hypogene origin."

---

[49] Oral communication.

[50] Work cited, p. 119.

[51] Work cited, pp. 119-20.

[52] Probably from 0.2 to 0.3 ounces of gold and 6 to 11 ounces of silver per ton.

When visited in January, 1934, the present owners were extending an adit tunnel that had been driven for 1,400 feet by Major Pickrell and was designed to intersect the vein at a depth of 300 feet below the deepest winze.

SENATOR MINE

The Senator mine, near the head of Hassayampa Creek, is accessible from Prescott by the Senator Highway. This deposit was opened by a short tunnel prior to 1871[53] and was worked mainly from 1883 to 1899. It has been held by the Phelps Dodge Corporation since 1889. Intermittent operations by lessees have yielded several cars of low-grade ore containing copper, gold, and silver.

Lindgren[54] has described the deposit as follows: "It is principally a gold property consisting of several parallel veins striking north-northeast. Among them are the Senator vein, carrying lead-zinc ores only and containing mostly gold with some silver; the Ten Spot vein, which carries mainly pyrite; the Tredwell vein, carrying heavy pyrite with specularite and gold; and the Snoozer vein, carrying copper ores with specularite. The shipping ore yielded $30 in gold and silver to the ton. There is a small mill on the property in which the ore from the Senator shoot was worked. The total production is said to be about $530,000 net, almost all of which came from the Senator ore shoot. Of this about $330,000 was extracted since 1890.

"Most of the veins crop out near the loop in the wagon road, which here ascends the Mount Union pass, at an altitude of about 7,000 feet. The Senator veins cut across the road a short distance below the Mount Union pass (altitude 7,188 feet) and also crop out on the ridge a short distance to the west. Here the main shaft was sunk 635 feet deep to the tunnel level (altitude about 6,500 feet). Below this tunnel, which extends eastward to the Cash mine, the shaft is continued for 200 feet, giving a total depth of 835 feet.

"The geology is complicated. There are lenses of diorite, amphibolite, and Yavapai conglomeratic schist, all traveised by dikes of rhyolite-porphyry. One dike of this kind, about 40 feet wide, cuts the Yavapai schist along the road a short distance below the Mount Union pass. A vein striking N. 40° E. from which a shipment was recently made crosses near the same place. It is probably the Ten Spot vein. The ore here, as exposed in a tunnel of quartzose ore, carries pyrite, chalcopyrite, and specularite.

"According to information kindly given by J. S. Douglas, who operated the mine between 1891 and 1893 and from 1896 to 1899, there was only one profitable ore shoot in the Senator properties, that of the Senator vein. This shoot starts just to the west of the old Senator shaft on top of the hill and pitches south-

---

[53] Raymond, R. W., work cited.
[54] Work cited, pp. 120-21.

ward until on the tunnel level, 600 feet below the collar of the shaft, the center of the shoot is 450 feet south of the shaft. The shoot had a stope length of 250 feet and averaged 18 inches in width. The Senator shoot carried milling ore with free gold associated with quartz, pyrite, galena, and zinc blende. Magnetite occurred in the vein only north of the shoot, where it crosses Maple Gulch.

"The Ten Spot, Snoozer, and Tredwell veins contained low-grade shipping ore with magnetite, specularite, chalcopyrite, and gold, but in the Senator property, none of this material was extracted at a profit."

CASH MINE[55]

The Cash vein, about 1,000 feet east of the Senator outcrops, is probably the extension of one of the Senator veins. Prior to 1883, it was opened by three shallow shafts. It was extensively worked from 1900 to 1902 and has been reopened for short periods at various times since. Lindgren says: "The mine is developed by a shaft 400 feet deep and has a 10-stamp mill with plates and concentration. The value of the total production could not be ascertained.

"Amphobolite schist crops out on the road between the Senator veins and the Cash. The shaft dump shows Yavapai schist injected with diorite. Dikes of normal rhyolite-porphyry about 10 feet wide crop out along the road to the mine. The vein strikes N. 40° E. and dips 60° SE."

Jaggar and Palache[56] describe the vein, which was being worked in 1900, as follows:

"The ore body in this mine is in the form of a series of well-defined lenses that have a maximum thickness of 2½ feet and occur in sericite schist which is in places black and graphitic. The ore is rich in sulphides, chiefly galena, sphalerite, pyrite, and chalcopyrite, contains some tetrahedrite in quartz, and is characterized by comb and banded structure, the center of the vein being generally open and lined with beautiful crystals of all the vein minerals. A rich body of free gold ore was found in this mine at a depth of 200 feet from the surface."

Lindgren continues: "The ore seen on the dumps in 1922 contains predominating quartz with some calcite and more or less pyrite, sphalerite, and chalcopyrite. Some of the sphalerite is coated with covellite. On the main ore dump was noted banded ore of magnetite and pyrite like that in parts of the Senator mine.

"Although the underground workings could not be visited, it seems clear that there are here two different kinds of veins, one of which contains magnetite, specularite, and adularia, in addition

---

[55] Lindgren, W., work cited, pp. 121-22.
[56] U. S. Geol. Survey Geol. Atlas, Bradshaw Mountains folio (No. 126). 1905.

to pyrite, chalcopyrite, sphalerite, and galena, the place of the iron oxides in the succession being between pyrite and chalcopyrite . . .   Ore of a second type carries apparently only chalcopyrite, galena, and sphalerite."

BIG PINE MINE

The Big Pine mine is about a mile west of the Senator, on the south side of Hassayampa Creek.   Lindgren states that it "has four tunnels between altitudes of 7,000 and 7,400 feet and a 200-foot shaft with drifts 200 and 350 feet long.   There is a cyanide plant on the property, which apparently has had little production.   C. H. Dunning states that the vein occupies a fracture in quartz diorite and contains quartz and pyrite.   The oxidation is said to be deep, and the ore shoots long and irregular.   The vein strikes N. 35° W. and dips 70° NW.   The ore is reported to contain $9 to the ton in gold and silver, about half of each by value."

"TRAPSHOOTER" REILLY PROPERTY

The holdings of "Trapshooter" Reilly Gold Mines include the old Crook, Venezia, Starlight, and Mount Union mines, in Crooks Canyon.   These mines are in the vicinity of Venezia post office, which, via the Senator Highway, is 15 miles from Prescott.

During the early days, these mines were rather extensively worked in the oxidized zone, which was generally from 40 to less than 200 feet deep.   Later, sulphide ores were mined and treated in a 20-stamp amalgamation-concentration mill at Venezia and in the Mount Union mill.   Much of this work was done by the late J. B. Tomlinson and lessees.   From 1927 to about 1932, the property was held by the Westerner Gold-Lead Mining Company which, according to the U. S. Mineral Resources of 1930, milled about 100 tons of gold ore and shipped one car of lead-zinc ore. Since about 1932, the property has been held by "Trapshooter" Reilly Gold Mines.   This company shipped some bullion and concentrates, but suspended operations in September, 1933.

The total combined production of these mines probably does not exceed $500,000.[57]

This area is composed of ridges and canyons with altitudes of 6,200 to 7,400 feet.   It lies in the granite belt that extends southward through Mount Union and contains many dikes of rhyolitic to basic composition.

Starlight group: Lindgren[58] gives the following description: "The three claims of the Starlight group lie about half a mile north of Venezia, at an altitude of about 6,600 feet.   They were formerly owned by an English company . . .   The developments consist of three short tunnels with a vertical interval of 200 feet. In the upper tunnel the vein strikes N. 30° E. and dips 60° W.

---

[57] Lindgren, W., work cited, pp. 123-25.
[58] Work cited, pp. 123-24.

"The upper tunnel runs along the vein for 300 feet. The vein is 4 or 5 feet wide and carries several 6-inch streaks of heavy galena and zinc blende. These solid streaks are said to yield high assays in gold and silver. Some ore has been packed up to the Mount Union mill for treatment. There is said to be a shoot 270 feet in length along the tunnel.

"The gangue is quartz-filling with an indication of comb structure. In part the vein has been reopened and brecciated. Other gangue minerals are ankerite and fine-grained fluorite. Pyrite with coarse-grained black sphalerite and more or less galena make up the ore minerals. A dike of rhyolite-porphyry of normal appearance shows in some places along the vein. Some of the breccia is cemented by galena and sphalerite."

According to local reports, the present company milled about 150 tons of ore from this group.

Crook vein:[59] The Crook vein outcrops at an altitude of about 7,000 feet, a short distance northeast of the Venezia road. It is very persistent and is reported to be traceable in a S. 10° E. direction for more than a mile.

During the early days, according to local people, its oxidized zone was worked by open cuts, mainly with arrastres, to depths of generally less than 40 feet, over a length of approximately 4,700 feet. The mine was operated shortly prior to 1902 by the Pan American Mining Company, and in a small way by lessees for some time afterwards. Attempts to work the lower levels of the vein were not generally successful. In 1933, the present company mined some ore from here and started an adit tunnel on the supposed extension at Venezia. The total production is estimated at $250,000.

The vein strikes N. 10° W. and dips 75° W. It occurs in gneissoid granite and follows, on the hanging wall, a persistent 15-foot dike of dark-green rock. It has been opened by a long tunnel, 160 feet below the outcrop, and by a 100-foot shaft. As seen in the tunnel, the vein ranges in width from a narrow stringer to about 4 feet, and averages about 1½ feet. Its longest ore shoot is said to have been 60 feet long by 20 inches wide. The vein filling consists of coarse-grained grayish-white quartz with some ankerite and abundant pyrite, galena, and black sphalerite.

Ore from a 100-foot shaft near the main tunnel contained quartz veins 4 to 6 inches wide with sulphides partly altered to chalcocite. The country rock here is partly sericitized quartzose schist.

At Venezia, a tunnel has recently been run northward for about 200 feet on the supposed extension of the Crook vein. It shows a few stringers of white quartz with some ankerite and pyrite.

A short distance below Venezia, the vein was opened by a 200-foot shaft, but the enterprise apparently was unsuccessful. About

---

[59] Mainly abstracted from Lindgren, W., work cited, pp. 124-25.

a half mile farther south, some rich ore was mined from a short tunnel where the vein is about 2 feet wide.

Mount Union mine:[60] The Mount Union mine, a quarter mile southwest of the summit of Mount Union, at an altitude of 7,400 feet, is accessible by a road that branches eastward at the divide. In 1906, this mine yielded lead ore with gold and silver. A small Huntington mill was operated for a short time, but the total production of the mine was not large.

Here, the rock is Bradshaw granite, cut by many dikes of rhyolite-porphyry. Two veins are reported to occur on the property, one of them from 6 to 10 feet wide, but they do not show well at the surface. Material on the dump shows pyrite, galena, and sphalerite in quartz. According to local reports, the ore was of low grade and the gold difficult to amalgamate.

GOLDEN EAGLE MINES

The Golden Eagle property of eight claims, held by C. M. Zander and R. M. Hansen, is on a branch of Slate Creek, in the western portion of the Bradshaw Mountains. By road, it is 5 miles from the Senator Highway and from Prescott.

Part of this ground was located in 1880. Developments consist of about 2,000 feet of tunnels and two 100-foot shafts. Surface equipment includes a power plant and a 25-ton concentrator, equipped with a ball mill, amalgamation plates, and a table. The U. S. Mineral Resources state that a small production was made by the property in 1925 and 1926.

Here, the rock is Yavapai schist, intruded on the east by a mass of diorite. As most of the surface is covered by dense brush, surface exposures are poor. Work has been done on several veins which generally strike northeastward and dip nearly vertically SE. They typically consist of small lenses and scattered bunches of grayish-white quartz in shear zones. In places, the quartz contains small irregular masses and disseminations of pyrite and chalcopyrite. The gold occurs both in the sulphides, particularly the chalcopyrite, and in fractures in the quartz.

TURKEY CREEK DISTRICT

The Turkey Creek district is in the vicinity of upper Turkey Creek, south of Bigbug Mesa. This region is made up mainly of Yavapai schist of sedimentary origin, intruded on the east and west by Bradshaw granite and on the south by the monzonite-porphyry stock of Battle Flat.

During the early days, this district was a notable producer of silver and gold. Since 1906, it has yielded a few thousand tons of ore that contained lead, silver, gold, and copper.

One gold-quartz mine, the Cumberland, is described by Lindgren[61] as follows:

---

[60] Abstracted from Lindgren, W., work cited, p. 125.
[61] Lindgren, W., Ore deposits of the Jerome and Bradshaw Mountains quadrangles, Arizona: U. S. Geol. Survey Bull. 782, p. 150. 1926.

"The Cumberland mine is one mile north of Pine Flat and was last in operation about 1908. According to information obtained from T. Roach, of Pine Flat, this is a gold quartz vein striking a little east of north and dipping west. The shaft is 350 feet deep and there are short drifts. The shoot on the north is said to have continued to the lowest workings; it contains ore of $40 grade, which was treated in a 10-stamp mill at Pine Flat. The gold is reported as free and visible in places. The water level was found at a depth of 90 feet.

"It is probable that here, as in many other places, the conditions became unfavorable when the workings reached the sulphide ore below water level. There is said to be placer ground in the vicinity of the Cumberland mine."

BLACK CANYON DISTRICT

The Black Canyon district comprises an area about 18 miles long by 8 miles wide that extends, between the eastern foot of the Bradshaw Mountains and the Agua Fria River, from the vicinity of Cordes on the north to the Maricopa County boundary on the south.

Here, a northward-trending belt of sedimentary Yavapai schist, about 2 miles wide, is intruded on the east and west by Bradshaw granite and on the east by a northward-trending strip of diorite. These formations floor a former valley and hilly pediment that is covered on the east by volcanic rocks and has been deeply dissected by the southward-flowing, meandering drainage system of Black Canyon. The elevation of the district ranges from 2,000 to 4,000 feet above sea level.

Lindgren[62] groups the gold-bearing veins of the Black Canyon district into two classes: (1) Pre-Cambrian quartz veins, mainly in the vicinity of Bumblebee and Bland Hill; and (2) quartz veins which dip at low angles and appear to be genetically connected with younger dikes of rhyolite-porphyry.

According to figures compiled by V. C. Heikes, of the U. S. Geological Survey, the Black Canyon district from 1904 to 1924, inclusive, produced $131,848 worth of gold, silver, lead, and copper. Of this amount $71,132 was gold of which a small percentage came from placers.

The pre-Cambrian veins, according to Lindgren, are of glassy quartz with free gold and some sulphides. They have furnished most of the gold for the placers of the district. He mentions the Cleator property, at Turkey Creek, and the Nigger Brown, Blanchiana, and Gillespie mines, south of Bumblebee, as having been worked in a small way.

GOLDEN TURKEY MINE

The Golden Turkey mine, held by H. C. Mitchell and associates, is on the west side of Turkey Creek, near the eastern foot of the

---

[62] Work cited, p. 153.

Bradshaw Mountains, at an elevation of about 3,000 feet. Via the Black Canyon Highway, the property is about 15 miles from Mayer, the nearest railway shipping point.

Some years ago, a 100-foot shaft was sunk on the property, but no production was made until 1933 when the workings were extended, and more than 4,000 tons of ore were run through the Golden Belt mill. The monthly yield was approximately two cars of concentrates containing gold and silver together with some zinc and a little copper. Several shipments of smelting ore also were made.

Here, pre-Cambrian schist strikes northward and dips almost vertically. The vein, which strikes northeastward and ranges in dip from 30° to less than 10° SE., occupies a fissure zone that is probably due to thrust faulting. As exposed, the vein ranges from a few inches to more than a foot in width and in places forms a branching lode several feet wide. The vein filling consists of very coarsely crystalline, milky to clear, glassy quartz together with rather abundant irregular masses and disseminations of pyrite, galena, and sphalerite. In places, a little chalcopyrite is present. The gold accompanies the sulphides, particularly the pyrite. The vein walls show rather intense sericitization and silification and in places contain disseminated metacrysts of practically barren pyrite.

When visited in January, 1934, developments consisted of a 500-foot inclined shaft and approximately 2,000 feet of workings. Most of the ore mined has come from below the 350-foot level, particularly where the vein flattens in dip. According to Mr. Mitchell,[63] the oxidized zone extended to a depth of approximately 250 feet on the incline.

GOLDEN BELT MINE[64]

The Golden Belt mine, held by the Golden Belt Mines, Inc., is a few hundred feet north of the Golden Turkey. Its original location is reported to have been made in 1873 by George Zika. Production prior to 1916, according to the local press, amounted to several hundred tons of ore of which part was shipped and part was milled. The present operators obtained this ground in 1931 prior to which time a small mill had been built and a little development work done. Subsequently, the mill has been rehabilitated and operated. In 1931, 134 tons of concentrates from 1,345 tons of ore were produced, and 107 tons of smelting ore were shipped.[65]

[63] Oral communication.

[64] In part abstracted from Minton, D. C., Cost of equipping and developing a small gold mine in the Bradshaw Mountains quadrangle, Yavapai County, Arizona: U. S. Bureau of Mines Information Circular 6735. 1933.

[65] U. S. Bureau of Mines Mineral Resources for 1931, Part I, p. 410.

Operations were continued through 1932 and 1933. Minton[66] states that the net smelter returns for 1931 and part of 1932 amounted to $891 on 106 tons of crude ore and $8,523 on 285 tons of concentrates. During 1933, ore from the Golden Turkey mine also was treated in the mill.

The geology of this mine is similar to that of the Golden Turkey. The prevailing rocks consist of pre-Cambrian schist, intruded by dikes of siliceous to basic porphyry. The Golden Belt vein, which strikes approximately N. 60° E. and dips from 10° to 23° SE., occurs within the fissure zone of a probable thrust fault. Minton[67] states that the vein, as exposed, ranges from a few inches to about 3 feet in width and carries from $5 to $40 in gold and from one to 10 ounces of silver per ton. In the upper portions of the mine, above the water level, the vein contains oxides of iron and lead and some free gold. Below the water level or 50 feet below the surface, galena and pyrite predominate, and the gold occurs mainly in the galena. The silver is asserted to be decreasing with depth.

In January, 1934, the mine was developed by an irregular, inclined shaft, reported to be 800 feet in length, with several hundred feet of drifts and stopes. Minton states that the cost of sinking 150 feet of the incline amounted to $6.22 per foot. As the workings extend under Turkey Creek, considerable water has been encountered.

The Golden Belt flotation mill has a maximum capacity of 50 tons per day. Power is obtained from the Arizona Power Company.

SILVER CORD VEIN[68]

"About a mile southeast of Turkey Creek station is the Silver Cord vein, said to be traceable through six claims. It dips south or southeast at an angle of less than 20°. This vein contains both silver and gold, together with some pyrite, galena, and chalcopyrite. The deposit was operated for several years, and twenty or thirty carloads of shipping ore is said to have been produced. In 1912, it is reported, 224 tons was shipped containing $40.67 a ton in gold, silver, copper, and lead. Among the properties on the Silver Cord vein is the American Flag, or the Old Brooks mine. Mr. Cleator, of Turkey Creek station, states that he shipped about twelve carloads from this property. The first-class ore contained $75 to the ton; the second-class ore $18 or more."

The U. S. Mineral Resources state that a small production of ore containing silver, lead, gold, and copper was made by the Silver Cord mine in 1925, 1928, 1929, and 1930.

---

[66] Work cited, p. 10.
[67] Work cited, p. 3.
[68] Quoted from Lindgren, work cited, p. 158.

FRENCH LILY MINE[69]

"The French Lily property, owned by Grove Brothers, of Mayer, is another of these peculiar flat veins. It was not visited but appears to be 2 miles southwest of Turkey Creek station, probably in granite. This vein, which is about 2 feet wide, dips 30° N. and is developed by an incline 190 feet long, with 250 feet of drifts on the second level. The best ore is said to contain 50 per cent of zinc and 1½ ounces of gold to the ton. One carload has been shipped, and it is claimed that 800 tons has been developed. The ore shows a filled quartz vein with comb structure. A considerable amount of ore was shipped from this property in 1923."

RICHINBAR MINE

The Richinbar mine is at an altitude of 3,500 feet on the western brink of Agua Fria Canyon, about 4 miles east of Bumblebee and 9 miles by road from Cordes.

From about 1905 to 1908, the Richinbar Mines Company carried on extensive development work at this property, built a 20-stamp mill, and, according to local reports, mined some 8,000 tons of ore that contained about $6 per ton in gold together with a little silver. Lindgren[70] states that, prior to 1922, the mine was operated by three different companies and, in 1917, was unwatered and retimbered. Early in 1933, the Sterling Gold Mining Corporation obtained control of the property and, prior to May, 1934, made some surface improvements and carried on a little underground work. Eight men were employed.

Here, the Agua Fria River has carved a meandering canyon 1,000 feet deep through the mesa basalt and into the underlying pre-Cambrian complex of schist, diorite, and granite.

The mine is on a vein that strikes north, dips almost vertically westward, and occurs principally in schist. Its main or Zyke shaft is 500 feet deep and is reported to connect with several thousand feet of workings that extend northward. When visited in May, 1934, the workings below the 200-foot level were under water. The upper workings are reached by two other shafts north of the Zyke shaft.

As seen in the accessible workings, the gangue consists mainly of coarse, massive, glassy quartz with some tourmaline and carbonate. In the ore shoots, the quartz tends to be more or less cellular to platy. From the surface to the 200-foot level, the ore is mostly oxidized but locally contains pyrite, chalcopyrite, galena, and sphalerite. Most of the ore mined was from three irregularly lenticular vertical shoots. The largest stope seen, which is on the 140-foot level, was about 65 feet long by 55 feet high by 14 feet wide. Throughout its exposures, the vein pinches and swells abruptly.

---

[69] Quoted from Lindgren, work cited, p. 159.
[70] Work cited, p. 157.

The schist wall rock shows considerable sericitization and locally contains limonite pseudomorphs after pyrite. Lindgren regards the vein as pre-Cambrian in age.

BRADSHAW DISTRICT[71]

"There are a number of small veins on the west side of Tuscumbia Mountain, most of them in Bradshaw granite.

"The Buster, owned by Charles Swazey, is a north-south vein dipping west. There are three tunnels, the longest 400 feet on the vein. The vein, of doubtful type, is 6 inches to 4 feet wide. Some ore has been milled in a 2-stamp mill at the mine, and the grade is reported to be $20 a ton. There has been some production from this vein.

"The Cornucopia is a parallel vein, owned by M. Roland. It is reported to be 18 inches wide. A tunnel follows the vein for 350 feet. A few years ago 100 tons of $9 gold ore from this vein was milled. There is some molybdenite in the ore.

"The Mohawk, one mile north of the Buster, is developed by a 300-foot shaft. There has been some production of gold ore, which was reduced in a small mill at Hooper.

"In all these veins there is, besides free gold, a considerable amount of sulphides. The free gold is probably derived from the oxidation of the sulphides."

PINE GROVE DISTRICT[72]

The Pine Grove district occupies a few square miles north of Crown King and east of Towers Mountain, at elevations of 6,000 to 7,500 feet. It is accessible from the Black Canyon Highway by a road that, west of Cleator, is built on the old railroad bed. A less used road connects Crown King with Mayer, and another route leads, via Venezia, from Crown King to the Senator Highway and Prescott.

During the early days, many veins in this district were found to be very rich in silver and gold near the surface. The total production of the district is, however, estimated at less than $3,000,000.

Here, a northeastward-trending belt of schist is intruded on the east by granite, on the west by granite and diorite, and on the southwest by granodiorite. These rocks are cut by northeastward-trending dikes of granite and rhyolite-porphyry.

Three prominent groups of veins occur in the district. They strike north-northeastward, dip about 60° NW., and are generally less than 5 feet wide. The shoots, which tend to pitch northward, occur in the granodiorite and to a less extent in the harder rocks.

The vein gangue consists mainly of quartz with some ankerite and calcite. Pyrite, chalcopyrite, sphalerite, galena, and tetra-

---

[71] Quoted from Lindgren, W., work cited, p. 176.
[72] Abstracted from Lindgren, W., work cited, pp. 164-65.

hedrite occur in the quartz of the unoxidized zone. Although some free gold is present in the primary zone, most of the ore mined was partly oxidized. In places, this ore contains $100 worth of gold and silver per ton, but the primary ores are generally about one-tenth as rich. Either silver or gold may predominate, but the richer silver ore tends to be accompanied by abundant ankerite. The water level in the Crown King mine is 250 feet below the surface.

The veins appear to be genetically connected with either the granodiorite or the rhyolite-porphyry and are probably of Mesozoic or early Tertiary age.

CROWN KING GROUP

The Crown King group of claims, southeast of Towers Mountain, includes the Crown King vein which has yielded the most of any deposit in the district.

History:[73] During the early days, this deposit yielded rich gold ore from near the surface. From 1890 to 1895, the property was operated by the Crown King Mining Company. Their 10-stamp mill caught $10 to $12 per ton on the plates and made a lead concentrate that contained $150 to $350 in gold and silver per ton, besides a middling that carried 15 per cent of zinc and $12 in gold per ton. The middlings were stock-piled, and the concentrates were packed to Prescott, some 40 miles distant, at a cost of about $21.50 per ton. In 1895, the property was bonded to H. B. Chamberlin and Company, of Denver. The mine was operated until 1901 when it was closed by court order. During the 1890-1901 period, the ore shoot was followed to a depth of 650 feet, with an estimated production of $1,500,000 of which $200,000 was paid in dividends.

Transportation costs were greatly reduced by the extension, in 1899, of the Prescott & Eastern Railway into Mayer and further by the completion, in 1904, of the Bradshaw Railway from Mayer to Crown King.

During 1906 and 1907, the Crown King Mines Company worked the middlings pile and shipped concentrates containing gold, silver, zinc, iron and copper.

In 1909, the property was sold at receiver's sale to the Yavapai Consolidated Gold-Silver-Copper Company controlled by the Murphy estate, for $75,000. In 1916, lessees organized the Bradshaw Reduction Company which installed flotation equipment and made a few tons of concentrates from old middlings. Some ore is reported to have been blocked out above the 480-foot level, but was not mined.

In 1923, the mine was taken over by the Crown King Consolidated Mines, Inc. During the winter of 1926-1927, a flood wrecked the mill. The railway tracks from Middleton to Crown King were torn up during 1926 and 1927, and the grade has been util-

---

[73] Abstracted from unpublished notes of J. B. Tenney.

ized for a road. In 1933-1934, a 300-ton flotation-concentration mill was built.

The production of the mine since 1890 is estimated at $1,840,000.

Vein and workings:[74] The Crown King vein strikes north-northeast and dips 70° W. It cuts across the contact between granodiorite and sedimentary Yavapai schist which has been intruded by a dike of rhyolite-porphyry. The vein averages 2 feet in width and, towards the north, splits. It contains two gently northward-pitching ore shoots. The ore consists of quartz with abundant sphalerite and pyrite and some chalcopyrite and native gold. It is said to average 0.5 ounces of gold and 4 ounces of silver per ton.

The mine workings include a 480-foot shaft, a 913-foot tunnel 150 feet below the collar, and a 500-foot winze with five levels that each extend about 1,200 feet north and from 200 to 500 feet south. Water stands 250 feet below the tunnel level. According to local reports, all of the ore down to the third level has been extracted, but some ore remains between the third and fifth.

DEL PASCO GROUP[75]

"About 4,000 feet east of the Wildflower is the Del Pasco vein, an old-time property which was worked in the early days and which has yielded a considerable production. It is first mentioned in Raymond's report of 1874. The Del Pasco strikes north-northeast, like the other veins in this vicinity, and dips 70° W. The main workings are on the south side of the ridge, at an altitude of 6,300 feet. Another tunnel enters from the north slope and taps the vein at an altitude of 6,600 feet. The dump at the north tunnel showed . . . much sphalerite, pyrite, and galena. The ore is said to contain gold with little silver. An upper tunnel on the north slope at an altitude of 6,700 feet exposed a vein said to be a branch of the Del Pasco, called the Jackson Strata. This was worked in 1922 by Reoff & Carner, who were also operating an arrastre. This vein is 2 to 3 feet wide and occurs in diorite intruded by a dike of typical rhyolite-porphyry. The quartz is partly oxidized and is said to contain $40 a ton."

PHILADELPHIA MINE

This property, which is on the Nelson and Gladiator veins, northeast of the Crown King, has been described by Lindgren[76] as follows: "The War Eagle Gladiator is a persistent vein traceable on the surface for at least 5,000 feet and developed by shafts and tunnels. It has doubtless yielded a considerable production of gold from oxidized quartzose ore near the surface. The strike is N. 20° E.

"A short distance to the east of this fissure is the Nelson vein,

---

[74] Abstracted from Lindgren, W., work cited, pp. 168-69.
[75] Quoted from Lindgren, W., work cited, pp. 167-68.
[76] Work cited, p. 169.

also traceable for a considerable distance and striking N. 10° E.
It is exposed at the shaft 250 feet deep near the top of the ridge
at an altitude of 7,000 feet and has also been opened by the Philadelphia tunnel at 800 feet from the portal. The tunnel follows
this vein for 3,000 feet but without reaching ore. The same vein
is also opened by an upper tunnel 280 feet above the Philadelphia
adit. In these workings, however, the Nelson vein shows with a
filling 2 to 5 feet thick, consisting of ankerite, with scant sulphides; it dips generally 60° to 80° W., but in places steeply east.
If the Nelson vein is identical with the vein at the shaft 1,000
feet higher up on the ridge, a change has taken place in the filling, for here it shows a quartz gangue with gold and silver
amounting to $20 a ton.

"Further explorations have been made to find the Gladiator
vein on the tunnel level, but so far without definite result."

FAIRVIEW TUNNEL[77]

"The Fairview vein, now owned by Ed. Block, of Prescott, lies
high on the ridge 2 miles north of Crown King. The southerly
end line of the claim has an altitude of about 7,200 feet. This
vein is considered to be an extension of the Nelson vein. From
this locality the outcrops drop sharply, and the lower tunnel,
about 1,300 feet farther northward, is at an altitude of about 6,900
feet. This lower tunnel, which is 230 feet long, shows 3 to 4 feet
of oxidized vein matter with seams of iron-stained quartz. The
fissure strikes N. 40° E. and dips 70° W. . . . The quartz, which
is generally massive, with a few small druses, contains some
pyrite and chalcopyrite. A carload of ore sacked at the portal
was reported to contain $30 a ton, mostly in gold.

"The country rock is generally a black clay slate of probably
sedimentary origin, striking N. 30° E. and dipping steeply east.
In places it contains ledges of quartzite and also several narrow
dikes of light-colored granite porphyry of coarser texture than
the rhyolite-porphyry connected with so many of the ore deposits
in this region."

LINCOLN MINE[78]

The Lincoln mine is 2 miles north-northeast of Crown King,
at an elevation of about 7,000 feet. It was worked in 1902 and
from 1905 to 1908, with a reported production of $135,000 from
10,000 tons of ore. It was equipped with a 15-stamp amalgamation-concentration mill.

The vein, which occurs in schist near the granite contact, strikes
N. 18° E. and dips steeply westward. This schist contains tourmaline and, near the vein, shows alteration to sericite and ankeritic carbonates. According to a local report, the ore occurs in a

[77] Quoted from Lindgren, W., work cited, p. 170.
[78] Abstracted from Lindgren, W., work cited, pp. 170-71.

northward-pitching shoot 400 feet long by 5 to 6 feet wide. It consists of drusy quartz and ankerite with pyrite, chalcopyrite, sphalerite, galena, and tetrahedrite. The concentrates are reported to have contained 15 per cent of copper, one to 4 ounces of gold, and 10 to 24 ounces of silver per ton.

Workings include two tunnels and an old 115-foot shaft.

TIGER DISTRICT[79]

The Tiger district occupies a few square miles south of the Pine Grove district, on the Humbug side of the divide. Its production, which probably amounts to less than $2,000,000, has been mainly in silver. One mine, the Oro Belle and Gray Eagle, produced considerable gold ore.

ORO BELLE AND GRAY EAGLE MINE[80]

The Oro Belle and Gray Eagle mine and mill, on the southern slopes of Wasson Peak, at an elevation of 5,400 feet, are accessible by road from the Tiger mine.

The Oro Belle and Gray Eagle were worked to some extent during the early days and from 1900 to 1912, with a total estimated production of $700,000. From 1907 to 1909, inclusive, according to Weeds' Mines Handbook, the yield was 5,662 ounces of gold, 16,301 ounces of silver and 23,830 pounds of copper. Some work is reported to have been done in 1915 and 1916. The prevailing rock is highly metamorphosed sedimentary schist, intruded by fine-grained granite dikes.

The Oro Belle vein is reported to be small. It was worked, mainly by lessees, through eight tunnels, of which the longest extends 1,000 feet.

The Gray Eagle vein is opened by a 600-foot shaft about ¼ mile north of the mill. It is near the schist-granite contact, and the foot wall is in pegmatite. This vein is said to be from 2 to 15 feet wide. Its ore is reported to have carried one per cent of copper, one ounce of gold, and 2 ounces of silver per ton.

MINNEHAHA VICINITY

Minnehaha is southwest of Crown King and southeast of Wagoner, from which it is accessible by road. This area is composed mainly of Bradshaw granite, overlain in places by andesites. Of the mineral deposits, Lindgren[81] says: "The scant mineralization is in part pre-Cambrian, in part apparently connected with the extension of the Crown King belt of dikes of rhyolite-porphyry.

"The Button shaft (at the head of Minnehaha Creek) is 400 feet deep, and from it drifts extend about 650 feet northward and 100

[79] Abstracted from Lindgren, work cited, pp. 174-75.
[80] Abstracted from Lindgren, W., work cited, pp. 174-75.
[81] Work cited, pp. 177-78.

feet southward.   The work was done about 1900 . . .   The deposit
is a pre-Cambrian quartz vein with glassy quartz and a little
pyrite, chalcopyrite, galena, and sphalerite.   The wall rock of
amphibolite shows no sericitization.

"About 1½ miles south-southwest of the Button deposit is the
Boaz mine, also closed down for many years.   This is an east-west
quartz vein which the late F. E. Harrington opened about 1902
and equipped with a 20-stamp mill and cyanide plant.   It is said
to be a 'spotty' vein, 'frozen to the walls.'   From its upper parts
some ore containing about $20 a ton in gold is reported to have
been mined.   One shaft is 650 feet deep, and drifts amount to
over 2,500 feet."

During 1933 and early 1934, R. E. Logan carried on intermittent operations on the old Colossal group, 12 miles from
Wagoner.   A little production was made with a small amalgamation-concentration plant.   Similar operations were conducted by
J. B. Paxton on the old Brown group, about 5 miles southeast
of Wagoner.

HUMBUG DISTRICT

HUMBUG MINES[82]

Situation and history: The holdings of Humbug Gold Mines,
Inc., in the southwestern Bradshaw Mountains or Humbug district, consist of approximately 100 claims and include the Fogarty,
Queen, Little Annie, Heinie, Lind, and Columbia groups.   Humbug camp, at an elevation of 2,600 feet on Humbug Creek, is accessible by 9½ miles of road which branches eastward from the
Castle Hot Springs Highway at a point 22¼ miles from Morristown.

In this area, gold mining was carried on with the aid of arrastres as early as 1880.   From 1900 to 1905, C. E. Champie operated
a 4-stamp mill at Columbia, on Humbug Creek.   Some ore was
shipped but, during the early days when Yuma was the nearest
shipping point, operations were greatly hampered by the inaccessibility of the district.   After 1905, only small-scale, intermittent work was attempted until 1932 when the present operators started active development.   According to Mr. Elsing, test
shipments of 207 tons of ore, mined from surface cuts and tunnels
on numerous veins, averaged approximately 1½ ounces of gold
and 3½ ounces of silver per ton, together with 3½ per cent of
lead.   A 50-ton flotation and table concentrating mill was completed and put into operation early in 1934.   In February of that
year, about eighty men were employed on the property.   Water
for all purposes is pumped from a shallow well near the bed of
Humbug Creek which normally is a perennial stream.

---

[82] Acknowledgments are due M. J. Elsing, C. L. Orem and F. de L. Hyde,
of Humbug Gold Mines, Inc., for important information.

Topography and geology: This ground, which lies within the drainage area of Humbug Creek and its branches, Rockwall and Carpenter creeks, has been eroded into sharp ridges and alternating southward-trending canyons about 800 feet deep. The prevailingly accordant summits of the main ridges appear to represent dissected remnants of the early Tertiary, pre-lava pediment that extends south of Silver Mountain.

Within this area, the principal rocks consist of large bodies of mica schist, surrounded by granite and intruded by numerous dikes of pegmatite and rhyolitic to granitic porphyry. The schist, granite, and pegmatite are regarded as pre-Cambrian in age, and the porphyry as Mesozoic or Tertiary.

The schistosity and the dikes prevailingly strike northeastward. Considerable pre-mineral and post-mineral faulting, principally of northeastward strike, is evident. Post-mineral faults of great magnitude follow some of the main gulches.

Veins: The veins of the Humbug area occur within fault fissures, mainly of northeastward strike and steep northwestward dip. Their filling consists of massive to coarsely crystalline, grayish-white quartz, together with irregular masses, veinlets, and disseminations of fine to coarse-grained pyrite and galena. In places, arsenopyrite is abundant. A notable amount of sphalerite is reported in one vein.

Most of the gold is contained within the iron minerals. The galena is reported to carry a little gold and locally as much as 40 ounces of silver per ton. Some free gold occurs as irregular veinlets and particles within fractures and cavities in the quartz. In the completely oxidized zone, which is generally of shallow, irregular depth, all of the gold is free.

These veins range in width from less than an inch up to 3 feet or more and persist for remarkably long distances along the strike. One of them is traceable on the surface for more than 9,000 feet. The ore shoots, which have been found to range from a few feet to a few hundred feet in length, are reported to contain from 0.25 to 9 or more ounces of gold per ton.

The wall rocks have been extensively altered to coarse sericite. Such alteration, together with the vein structure, texture, and mineralogy, indicates deposition in the mesothermal zone. Not enough work has been done to reveal the structural features that determine their ore shoots. Apparently, the high-grade portions are less than a foot wide, but the altered condition of the country rock permits cheap mining by a lessee system. According to Mr. Orem, the total cost of drifting during preliminary development ranged from $1 to $4 per linear foot.

CASTLE CREEK DISTRICT

The Castle Creek district is in southern Yavapai County, in the vicinity of upper Castle Creek. It is accessible by unimproved roads that lead from Wickenburg, Wagoner, and the Castle Hot Springs highway.

This region is made up mainly of Yavapai schist and Bradshaw granite, locally intruded by dikes of diorite and rhyolite-porhpyry and largely mantled on the south by volcanic rocks. It has been deeply and intricately dissected by the southeastward-flowing drainage system of Castle Creek. As the elevation ranges from about 2,500 to generally less than 4,000 feet, the streams carry water only occasionally, and desert vegetation prevails.

The ore deposits, which occur only in the pre-Cambrian rocks, have been grouped by Lindgren[83] as follows: Pre-Cambrian gold-quartz veins, represented by the Golden Aster or Lehman deposit; post-Tertiary gold-copper veins, exemplified by the Swallow, Whipsaw, Jones, and Copperopolis properties; and lead veins. Lindgren states that the total production of the district, including rich ore shipped and ore treated in the Lehman and Whipsaw mills, probably amounts to less than $500,000.

GOLDEN ASTER OR LEHMAN MINE

The Golden Aster mine is at an altitude of about 4,200 feet on a granite ridge about 1⅛ miles north of Copperopolis. It is accessible by road and trail from Wagoner and by trail from Copperopolis.

During the early days, this deposit was owned by Gus Lehman. Later, it was acquired by E. C. Champie. Some ore was treated in a 5-stamp mill on Spring Creek, a short distance west of the mine. For the past several years, a small production has been made. About 40 tons of ore was shipped in 1932-1933.[84] Early in 1934, three men were employed.

Here, the prevailing rock is granite, with some inclusions of schist and dikes of pegmatite. The deposit consists of closely spaced, parallel, branching veins that strike N. 10° W., dip 25° W., and range from a few inches to 2 feet in width. Their filling is massive, glassy quartz with limonite and a little tourmaline. Coarse free gold is locally present.

The mine is opened by about 1,000 feet of tunnels and raises.

BLACK ROCK DISTRICT

ORO GRANDE MINE

The Oro Grande property of fifteen claims in southern Yavapai County is about a mile east of the Hassayampa River and 4½ miles by road north of Wickenburg.

This deposit is reported to have been prospected in a small way for copper and silver during the seventies. In 1900, it was acquired by G. B. Upton and associates who, during the following three years, sank a 340-foot shaft, did a few thousand feet of de-

---

[83] Lindgren, W., work cited, pp. 183-84.
[84] Oral communication from Joe Stockdale.

velopment work, and built a 10-stamp mill. In 1904, they milled about 8,600 tons of ore which yielded an average of $5.32 worth of gold per ton.[85] Water for this operation was pumped from the Hassayampa River. The present company, Oro Grande Consolidated Mines, was incorporated in 1929. When visited in February, 1934, the workings below the 240-foot level were under water.

The mine is in a moderately hilly area composed of altered diorite with lenses of basic schist and irregular dikes of aplite and pegmatite. At the mine, a fractured to brecciated zone, in places more than 100 feet wide, forms the southeastern wall of a nearly vertical fault that strikes N. 37° E. As seen in the outcrops, portions of this zone are cut transversely by narrow, irregular veins and stringers of white quartz in the walls of which are developed pseudomorphs of limonite after pyrite. Drywashers have recovered a little coarse placer gold from this vicinity.

The principal ore body found in the mine occurred near the southwestern exposure of the brecciated zone. More or less discontinuous stopes between the surface and the 200-foot level indicate that it was of irregular shape and several tens of feet in maximum width. This ore consists of brecciated country rock cemented with brown to black limonite, calcite, and coarsely crystalline, glassy quartz. The gold occurs as ragged particles, mainly in the limonite associated with the quartz. The silver content is reported to be low except where local concentrations of oxidized copper minerals are present.

According to Mr. Upton, a considerable amount of low-grade gold-bearing material was exposed by the drifts that explored the breccia zone east and northeast of this stoped area.

GOLD BAR OR O'BRIEN MINE

The Gold Bar or O'Brien mine is 15 miles by road northeast of Wickenburg and 2.7 miles northeast of Constellation.

This deposit was located in 1888 by J. Mahoney. About 1901, the Saginaw Lumber Company erected a 10-stamp mill on the property and is reported to have treated 4,000 tons of ore that yielded about $60,000.[86] In 1907-1908, the Interior Mining and Trust Company is reported to have mined the ore body from the surface to the 385-foot level on the incline. This company erected a 100-ton mill, equipped with stamps, amalgamation plates, tables, and vanners. Heikes states that the 1907 production amounted to $33,402 in bullion and concentrates. These concentrates averaged, per ton, 2 ounces of gold, 3 ounces of silver, 49 per cent of iron, 15 per cent of silica, and 15 per cent of sulphur.[87] He also states that, in 1908, $91,749 worth of gold came from the Black Rock district of which the largest producer was the Interior Mining

---

[85] Oral communication from G. B. Upton.
[86] Oral communication from Ward Twichell.
[87] U. S. Geol. Survey, Mineral Resources, 1907, Pt. I, pp. 182-83.

and Trust Company.[88]   About 1915, the company was reorganized as the Gold Bar Mining Company and a vertical shaft was sunk to the 700-foot level.   In February, 1934, the property was under the trusteeship of the Commonwealth Trust Company, of Pittsburg, and was being worked in a small way by lessees.

This region has been deeply dissected by northward-flowing tributaries of Hassayampa Creek.   The principal rock is medium-grained granite, with some inclusions of schist.   It is intruded by pegmatite, granite-porphyry, and basic dikes.   Fissuring in N. 70° E. and S. 30° E. directions is evident.   The vein, which outcrops on the western side of O'Brien Gulch, at an altitude of 3,400 feet, occurs within a fissure zone that strikes N. 70° E. and dips 30° NW.   Its filling consists of coarsely crystalline, glassy, grayish-white quartz.   In places, the quartz from the oxidized zone is rather cellular with cavities that contain abundant hematite and limonite formed from pyrite.   Pyrite is present in the deeper workings.   The gold occurs as fine to mediumly coarse particles, both in the quartz and with the iron minerals.   The wall rock shows intense sericitization.

The mine workings indicate that the ore shoot was a chimney that measured about 40 by 50 feet in cross-section at the surface and plunged 30° SW.

GROOM MINE

The Groom or Milevore Copper Company mine is near the head of Amazon Gulch, about 7 miles by road east of the Constellation road about 16 miles northeast of Wickenburg.

A few years ago, this property was prospected for copper by means of a 500-foot shaft and considerable diamond drilling.

This region is composed largely of granite which has been complexly faulted and deeply oxidized.   The principal mineralization is in two nearly vertical veins, some 250 feet apart, that strike N. 20° W. and are traceable on the surface for more than a mile.   Their filling consists mainly of hematite, limonite, fine-grained quartz, and locally abundant chrysocolla, malachite, and copper pitch.   The eastern vein, which shows the stronger mineralization, is generally from 3 to 6 or more feet wide but pinches out in places.   It is accompanied by an altered dike, apparently of intermediate composition, and has been cut by numerous transverse faults.

When visited in February, 1934, R. L. Beals and associates were obtaining a little siliceous gold ore from shallow workings on the eastern vein.   This ore was being treated in a 30-ton plant equipped with a ball-mill, classifier, amalgamation plates, and corduroy-surfaced tables.   Rather fine grinding is required.   According to Mr. Beals, the ratio of gold to silver in the bullion is about 7 to 3.   Water is obtained at shallow depths.

---

[88] Work cited, 1908, Pt. I, p. 310.

ARIZONA COPPER BELT MINING COMPANY

The Arizona Copper Belt Mining Company, whose property is a short distance northwest of Constellation, was formed in 1906.[89] The U. S. Mineral Resources state that it made a small production of gold-bearing copper ore in 1912. During the past few years, attempts have been made to develop gold ore on the property. A 25-ton Gibson plant, equipped with a rod mill, amalgamator, table, and flotation unit, was nearing completion in March, 1934, when operations were suspended.

The older openings include two shafts, reported to be 250 and 300 feet deep, together with several hundred feet of workings. At the time visited, one of these shafts was caved, and the other was full of water to the 100-foot level. More recent openings include a 200-foot inclined shaft with a few hundred feet of drifts. At this shaft, a vein in fine-grained, sheeted granite strikes S. 65° W., dips 45° N., and ranges in thickness from less than one foot to about 6 feet. Its filling consists typically of brecciated massive gray quartz, cemented with yellowish limonite.

WHITE PICACHO DISTRICT

GOLDEN SLIPPER MINE

The Golden Slipper property of seven claims, in the Wickenburg Mountains or White Picacho district of Southern Yavapai County, is accessible by one mile of road which branches north from the Castle Hot Springs Highway at a point 10 miles from Morristown.

This deposit was discovered many years ago but has been worked only intermittently and in a small way. According to data kindly supplied by P. C. Benedict, the known record of production within the last few years prior to March 15, 1934, is as follows:

Tons of Ore	Gross Value	
0.5 approx.	$5,000)	
30.0	600)	
30.0	1,000)	With gold at $20.67 per ounce
30.0	600)	
30.0	315)	
38.0	430)	
100.0	2,000)	
50.0	1,750)	With gold at $25 to $35 per ounce
50.0	3,300)	

The mine is on a hilly pediment of platy to schistose rhyolite whose fracturing strikes northwestward and dips almost vertically. Detailed mapping by P. C. Benedict shows that the ore so far found is associated with a southeastward-trending syncline. The vein occurs within a brecciated fault zone that strikes southeastward and dips 20° or more NE. Its filling consists of brec-

---

[89] Weed, W. H., The Copper Handbook, vol. 11, p. 72, 1912-1913.

ciated rhyolite with irregular stringers of grayish to pale-greenish quartz. This quartz is fine-grained and dense, with a few druses. Some iron oxide and a little local copper stain are present. The gold is finely divided and occurs mainly in the grayish quartz.

In February, 1934, the Golden Slipper mine was being operated by S. Nixon and C. W. Nelson, lessees. The workings included a 25-foot shaft, with about 300 feet of drifts and stopes. The principal known ore shoot was from 1½ to 3 feet thick by about 25 feet wide and plunged with the dip of the vein but died out within a few tens of feet. For some distance above this ore body, the rhyolite is strongly marked with yellowish-brown iron stain. Alteration near the vein consists mainly of sericitization and silicification. This deposit is evidently of the epithermal type.

WEAVER DISTRICT

The Weaver district, of southern Yavapai County, is at the southwestern margin of the Weaver Mountains, in the vicinity of Octave and Stanton. This range or plateau, whose dissected front rises abruptly for more than 2,000 feet, is made up mainly of granite, diorite, and schist, intruded by numerous aplites, pegmatites, and basic dikes, and overlain in places by lavas. The southwestern foot of the range is separated by a narrow pediment from open desert plains.

The principal gold-bearing veins of the Weaver district occur within gently north-northwestward dipping fault zones and are of the mesothermal, quartz-pyrite-galena type.

OCTAVE MINE[90]

Situation: The Octave mine, at Octave, is accessible by about 10 miles of road which leads eastward from Congress Junction.

History and production: This deposit probably became known in the sixties, shortly after the discovery of the Rich Hill placers, but, as a large part of the gold was not free-milling, little work was done upon it until the advent of the cyanide process. During the late nineties, according to local reports, a group of eight men purchased the property and organized the Octave Gold Mining Company. Between 1900 and 1905, the vein was mined to a depth of about 2,000 feet on the incline and for a maximum length of 2,000 feet along the strike. This ore was treated in a 40-stamp mill equipped for amalgamation, table concentration, and cyanidation. The total gold and silver production during this period is reported to have been worth nearly $2,000,000. In 1907, the property was bought by a stock company that built an electric power plant at Wickenburg, 11 miles away, and electrified the mine and mill. This company, however, failed to work the mine at a profit and ceased operations in 1912. In 1918, a group of the stockholders organized the Octave Mines Company, with H. C.

---

[90] Acknowledgments are due B. R. Hatcher, Carl G. Barth, Jr., M. E. Pratt, and A. E. Ring for much information upon the Octave region.

Gibbs, of Boston, as president, and carried on development of the Joker workings until 1922. In 1928, the Arizona Eastern Gold Mining Company, Inc., was organized. This company operated a 50-ton flotation plant on ore already blocked out in the Joker workings. After obtaining approximately $90,000 worth of concentrates from 9,100 tons of ore, operations were suspended in 1930. This ore contained nearly equal proportions by weight of gold and silver.[91]

When visited in February, 1934, the mine was idle, and the levels below the Joker workings were largely under water.

Topography and geology: The Octave mine is at an elevation of 3,300 feet on a narrow pediment at the southwestern base of the Weaver Mountains and 2 miles south of Rich Hill. This area is dissected by several southward-trending arroyos of which Weaver Creek, ¼ mile west of Octave, is the largest. Water for all purposes is brought through 7 miles of pipe line from Antelope Spring.

The pediment and mountains immediately northeast of Octave consist mainly of grayish quartz diorite with included lenses of schist and northeastward-trending dikes of pegmatite, aplite, and altered, fine-grained basic rock. Examined microscopically in thin section, the basic rock is seen to be an altered diabase. The pegmatite and aplite dikes are earlier than the veins, and the diabase dikes cuts the veins. As shown by Mr. Barth's geologic and relief model of the district, the dike systems are rather complex.

Mineral deposit: The main Octave vein occurs within a fault fissure that strikes N. 70° E. and dips from 20° to 30° NW. Cleavage in its gouge indicates that the fault was of reverse character. This fissure, which is traceable on the surface for several thousand feet, has been intersected by three or four systems of post-vein faults of generally less than 100 feet displacement.

As seen in the Joker workings, the vein-filling consists of coarse-textured, massive to laminated, grayish-white quartz together with irregular masses, disseminations, and bands of fine-grained pyrite, galena, and sparse chalcopyrite. Most of the gold is contained within the sulphides, particularly the galena, and comparatively little is reported to occur free. According to Mr. Pratt, the pure galena generally assays more than 100 ounces of gold per ton, the chalcopyrite 8 to 25 ounces, and the pyrite from 3½ to 7 ounces. A band, from a few inches to a foot thick, of barren, glassy, grayish quartz commonly follows one of the vein walls.

The width of the vein ranges from a few inches to 5 feet and averages approximately 2¼ feet. In places, it narrows abruptly and forms branching stringers. Locally, a parallel vein occurs

---

[91] Oral communication from H. C. Gibbs and M. E. Pratt.

about 25 feet away from the main vein, but its relations have not been determined. Wall-rock alteration along the Octave veins consist of sericitization, silicification, and carbonatization.

The Joker shaft is about 1,100 feet deep on the incline and connects with a few thousand feet of drifts. These workings found, below the 600-foot level, two ore bodies of which the larger was followed for a horizontal length of about 700 feet by a maximum height of 200 feet.

The old Octave workings, which extended to a depth of 2,000 feet on the incline, included four shafts with several thousand feet of drifts and extensive stopes on three ore shoots. A sketch section of these workings has been published by Nevius.[92]

OTHER PROPERTIES

The *Alvarado* or old *Planet Saturn* mine, at the mouth of Fool's Gulch, about 5 miles northeast of Congress Junction, is reported to have made a small production, prior to 1905, from two inclined shafts that were approximately 800 feet deep. Its surface equipment was dismantled in 1920.

According to Carl G. Barth, Jr.,[93] a few pockets of high-grade gold ore were mined from a vein in schist on the *Leviathan* property, at Stanton. He also states that, in 1912, a narrow vein on the *Rincon* property, about a mile farther northwest, was worked to a depth of 1,100 feet on the incline.

Two cars of sorted ore are reported to have been shipped during the past 2 years from a narrow, gently northward-dipping vein on the property of the *Rich Hill Consolidated Gold Mines, Inc.*, at the southeastern end of Rich Hill.

The *John Sloan* property, on the steep southwestern slope of Rich Hill, has recently produced a little gold from a narrow quartz-pyrite-galena vein in granite.

The *Bee Hive* or *Zieger* property is about 2 miles northeast of Octave where a northwestward-trending shear zone intersects the supposed continuation of the Octave fissure. The principal workings are east of the quartz diorite, in schist which is intruded by numerous dikes of aplite and pegmatite. Only a few thin veinlets of quartz are present, but some flat nuggets and flaky gold were found in the sheared schist. A small production was made with pans and arrastres, but approximately ½ mile of tunnel and two 90-foot shafts are reported to have found very little ore.

The *Bishop* property, about a mile south of the Zieger, is reported by Mr. Barth to have produced some gold ore from a narrow, gently northward-dipping vein.

---

[92] Nevius, J. N., Resuscitation of the Octave gold mine: Min. and Sci. Press, vol. 123, pp. 122-24, 1921.

[93] Oral communication.

MARTINEZ DISTRICT

The Martinez district, of southern Yavapai County, is at the southwestern margin of Date Creek Mountains, west of the Weaver district and a few miles northwest of Congress Junction. This westward-trending range, whose steep southern slope rises for 2,000 feet above the adjacent plain, is separated from the Weaver Mountains by the canyons of Martinez and Date Creeks. It is made up mainly of coarse-grained granite, intruded by pegmatites, aplites, and basic dikes.

The principal gold-bearing veins of the Martinez district occur within northward-dipping fault zones. They represent a mesothermal type and contain coarse-textured quartz together with pyrite and some galena.

Raymond[94] stated that several gold deposits were discovered in the Martinez district early in 1870. As comparatively little of their gold, even near the surface, was free milling, only the richest portions could be worked at a profit until the introduction of the cyanide process in 1895.

CONGRESS MINE[95]

Situation: The Congress mine is 3 miles by road northwest of Congress Junction, a station on the Santa Fe Railway and U. S. Highway 89.

History: The Congress deposit, according to Staunton, was located by Dennis May who, about 1887, sold it to "Diamond Joe" Reynolds for $65,000. Staunton, who had supervision of the property from 1894 to 1910, continues: "Reynolds developed the property to some extent and built a 20-stamp mill with Frue vanner tables for concentration. No amalgamating plates were used, as there was practically no free gold . . . The surface ores were much oxidized, in spite of which no saving of consequence could be made by amalgamation or by concentration. The cyanide process was in its infancy then and little known, so that it was commonly said of the Congress mine in its early history that though it showed much good ore, there was no known method of extraction. The finding of sulphides by sinking solved the problem to a certain extent, as such ores were amenable to concentration and the concentrates could be shipped to custom smelters. This furnished the means to profitable operation, but the crude methods employed at that time—fine crushing by stamps followed by simple unclassified concentration on Frue vanners— necessarily resulted in high tailing losses on account of the large amount of sliming that took place. Flotation, as practiced today,

94 Raymond, R. W., Statistics of mines and mining in the states and territories west of the Rocky Mountains, p. 255. Washington, 1872.

95 Staunton, W. F., Ore possibilities at the Congress mine: Eng. and Mining Jour., vol. 122, No. 20, pp. 769-71. Nov. 13, 1926.

was then unknown.   Fortunately the tailings from the early operations were saved and were re-treated later by cyanide with
good extraction.

"The property was operated from March, 1889, to August, 1891,
when owing to the death of Mr. Reynolds, and to await the construction of the Santa Fe, Prescott and Phoenix R. R., active
operation was suspended except for a certain development work
and the enlargement of the mill from 20 to 40 stamps with the
necessary additional Frue vanners.   The No. 2 shaft had been
sunk to a depth on the vein of 1,000 feet, but no stoping had been
done below the 650-foot level.

"In March, 1894, new interests acquired control of the company,
the name of which then was the Congress Gold Company, with
E. B. Gage, president, and active operations were resumed, continuing thereafter until the end of 1910."   Water was pumped
from a well in Date Creek, 8 miles distant.

Blake[96] states that, in 1895, a plant was built to treat the tailings
by light roasting and cyanidation.

In 1901, the company was reorganized as the Congress Consolidated Mines Company, Ltd. About four hundred and fifty men
were employed in the mine and mill.

Production: Staunton gives the total recorded production as
follows:

PRODUCTION OF CONGRESS MINE

	Tons	Net returns
March 3, 1889 to Aug. 31, 1891, ore shipped	1,129.4	$ 155,652.29
Sept. 26, 1889 to Jan. 28, 1891, concentrates ship- ped	2,500.8	335,308.87
June 3, 1891 to Aug. 31, 1891, concentrates shipped	1,062.8	101,113.73
March, 1894 to Dec., 1910, concentrates and ore shipped	3,661.0	4,259,571.30
March, 1894 to Dec., 1910, cyanide bullion shipped		2,797,851.45
Total		$7,649,497.64

TONS OF ORE MINED FROM THE VEINS

	From Congress, Tons	From Niagara, Tons	From Queen of the Hills, Tons	Total Tons
March 3, 1889 to Aug. 31, 1891	71,129			71,129
March 1, 1894 to Dec. 31, 1910	307,863	293,215	20,125	621,203
Total tons	379,022	293,215	20,125	692,332

"The recorded production of gold and silver in shipments shows
a total of 388,477 ounces of gold and 345,598 ounces of silver. As
this came from 692,332 tons of ore, a recovery is indicated of
$11.81 a ton, gold being figured at $20.67 and silver at 60c per
ounce. Average tailing assays were about $1.20, which indicates

---

[96] Blake, Wm. P., in Rep't. of Gov. of Ariz. for 1898, pp. 248-51.

a gross average value of all ore mined of $13.01." A total of 687,542 tons of ore was milled.

Except for a few attempts to work the dumps, the property has remained more or less dormant since 1910. The surface equipment and the 4-mile railroad leading to Congress Junction were dismantled about 1920. During 1923, 1925, 1926, and 1931, lessees shipped a few cars of ore from the property. In 1928, several thousand tons of mine dump material were treated by combined gravity concentration and flotation. In 1931, the Southwest Metal Extraction Corporation treated about 10,000 tons of the old mill tailings dump by cyanidation. Early in 1934, the Illinois Mining Corporation began treating material from the mine waste dumps in a 150-ton plant equipped with a ball mill, tables, and cyanide tanks, but suspended these operations after several weeks' run. In May, 1934, this company was reconditioning No. 2 shaft and having a geophysical survey made.

Geology: The Congress mine is at an elevation of 3,400 feet at the southern base of the eastern end of the Date Creek Mountains. This portion of the range consists essentially of coarse-grained biotite granite, intruded by aplite, pegmatite, and greenstone dikes. These greenstone dikes, where unaffected by superficial alteration, are fine-grained, dense, and greenish black. Examined microscopically in thin section, the rock is seen to consist of a finely divided aggregate of altered hornblende together with abundant calcite and quartz. Scattered grains of magnetite and pyrite are present. Staunton states that the following analysis of an average specimen of the greenstone was reported from the Sheffield Scientific School: SiO_2, 52.20 per cent; Al_2O_3, 13.40; FeO, 9.75; MnO, 1.90; CaO, 9.60; and MgO, 1.16.

Veins and workings: This property contains several gold-bearing quartz veins of which the Congress, Niagara, and Queen of the Hills have been of particular economic importance. These veins occur within fault fissures which strike generally westward and dip northward.

The Congress fissure dips 20° to 30° N. and occurs largely within a greenstone dike that is about 15 feet thick. The vein filling consists of coarse-textured, massive, grayish-white quartz together with irregular masses, bands, and disseminations of fine-grained iron sulphide. Staunton regarded the iron sulphide as marcasite, but microscopic examination in polished section proves it to be pyrite. Galena is rare, and very little free gold has been reported.

Although the vein as a whole follows an irregular course within the greenstone dike, the ore-bearing portions are generally flat lenses near the footwall. These lenses commonly terminate as stringers. Staunton says: "Although the Congress vein is continuous and well defined for a mile or more to the west of the mine workings and shows both the characteristic quartz and sulphides, the pay ore was practically confined to a shoot in the vein pitching to the northwest and coinciding closely with the intersection of one of the fissure veins in the granite. The granite vein is faulted by the Congress vein so that the intersection is obscure in the mine workings. The portion of the granite vein in the hanging wall of the Congress carried bodies of pay ore.

"The Congress pay shoot varied greatly in length on different levels, being longest on the 650-ft. level, where it was stoped continuously for 1,800 feet. The average thickness of pay ore was less than 3 feet. Several pinches were met in following the vein down, the most serious being at the 1,700-ft. level, where there was no stoping ground. On the theory that if pay ore existed below that point it would probably be found on the general line of trend of the ore shoot above, a deep prospecting winze was sunk from the 1,700-ft. level, in the vein but with a northwesterly pitch corresponding to the established trend of the pay ore in the upper workings. This winze was sunk 1,000 feet and bore out fully the theory upon which it was projected, the pay ore coming in again as good as ever after a few hundred feet of lean ground.

"The 3,900-ft. level was the deepest point at which any considerable amount of development was done. For several levels above, there had been a gradual pinching of the pay shoot, which became small and irregular, although retaining its mineralogical characteristics and the small amount of sulphides which remained still showing the characteristically high gold contents, about 7 ounces per ton. The conditions were similar to those existing at other horizons in the mine where persistent deeper work had been rewarded by expansion of the ore shoot to normal size.

"There are other veins entirely in the granite and unaccompanied by the greenstone so characteristic of the Congress vein. These strike east and west, but dip more steeply, from 40° to 50°. The development of quartz is more extensive than in the Congress vein and the average grade is lower. One of these veins, the Niagara, carried large bodies of ore of commercial grade to a depth of 2,000 feet. A characteristic of these all-granite veins is the presence of a small amount of galena and higher silver content.

"Minor faulting is in evidence throughout the mine workings and there has been considerable relative movement of the walls of the Congress vein resulting in local crumpling of the greenstone. The mine workings terminate to the east against a heavy fault, beyond which the vein has not been definitely located. This fault cuts off both the Congress and Niagara veins.

"The mines were practically dry down to the deepest point reached, 4,000 feet on the Congress vein at an approximate inclination of 25° from the horizontal, the small amount of surface water which found its way in being easily handled by bailing tanks in the shafts. No mine pumps were ever put in or needed.

"Seven shafts were sunk, all of them inclines following the veins. Three of them were on the Congress vein, designated as No. 1 (1,100 ft.), No. 2 (1,700 ft.) and No. 3 (4,000 ft.). On the Niagara vein three shafts were also sunk, No. 4 (1,000 ft.), No. 5 (2,050 ft.), and No. 6 (1,800 ft.). On the Queen of the Hills vein, one shaft was sunk to a depth of 200 feet below the tunnel level."

When visited in February, 1934, the mine workings below the 1,350-foot level of No. 2 shaft were under water.

CONGRESS EXTENSION MINE

The Congress Extension mine, about ½ mile west of the Congress mine, is reported to have been operated for about three years during the nineties by T. Carrigan and associates who sank a 950-foot inclined shaft and opened nine levels, each of which extends for approximately 100 feet east and west of the shaft. The Herskowitz family obtained the property in 1910 and, since about 1923, has made yearly shipments of 35 to 80 tons of ore that contained from $8.50 to $17.50 worth of gold per ton.[97]

When visited in May, 1934, the shaft contained water to the 140-ft. level. Where seen in the upper workings, the vein strikes westward, dips 35° to 40° N., and occurs in granite. A stope on the first level east of the shaft indicates that the ore there occurred as a flat lens about 3 feet thick. The vein material consists mainly of massive, locally shattered white quartz. Specimens of this quartz seen on the dump contain irregular bunches and disseminations of pyrite and a little chalcopyrite. The oxidized vein material contains abundant limonite and some copper stain.

As the surface exposures are terminated on the east by alluvium, the relations of this vein to the Congress and Niagara veins are unknown.

CHAPTER III—MOHAVE COUNTY

GENERAL GEOGRAPHY

Mohave County, as shown on Figure 3, comprises an irregular area about 190 miles long from north to south by 85 miles wide. Its north-northeastern portion is within the deeply dissected Plateau Province and rises to an altitude of more than 6,000 feet. The rest of the county consists of north-northwestward-trending fault-block mountain ranges and valleys. The largest of these ranges is the Black Mountains, near the Colorado River. This range is nearly 100 miles long by a maximum of 20 miles wide and attains an altitude of more than 5,000 feet.

The region is drained by the Colorado River, as indicated on Figure 3.

---

[97] Oral communication from H. Herskowitz

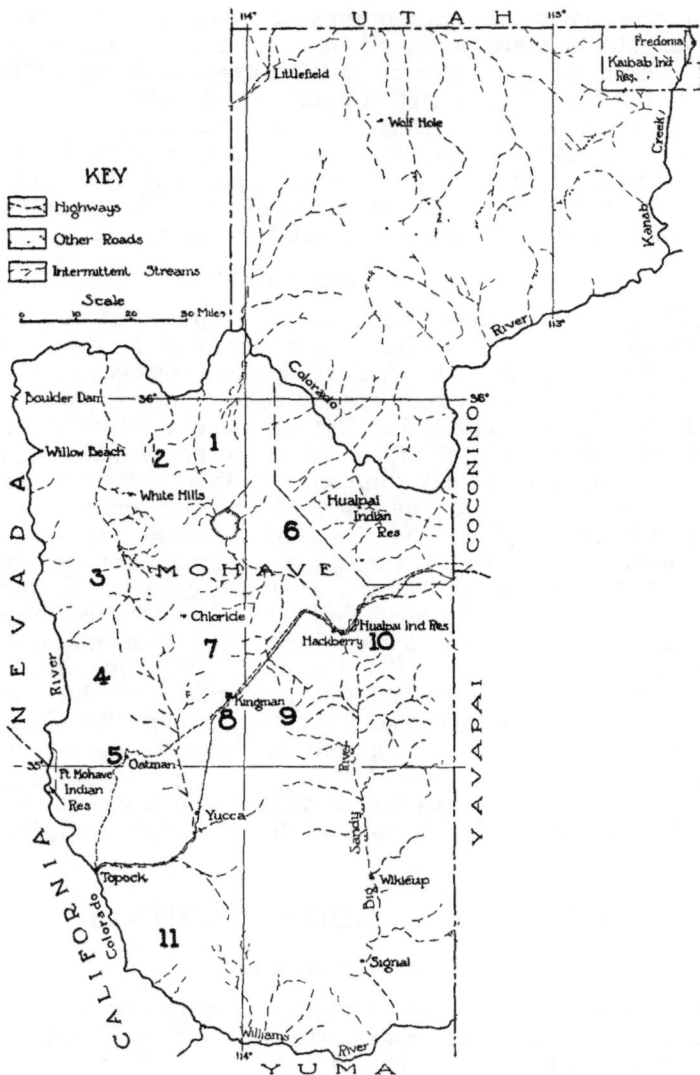

Figure 3.—Map showing location of lode gold districts in Mohave County.

1 Lost Basin
2 Gold Basin
3 Northern Black Mountains
    (Weaver, Pilgrim)
4 Union Pass
5 Oatman
6 Music Mountain
7 Cerbat Mountains (Chloride,
    Mineral Park, Cerbat, Walla-
    pai)
8 McConnico
9 Maynard
10 Cottonwood
11 Chemehuevis (Gold Wing)

The highest and lowest temperatures on record for Kingman (altitude 3,326 feet) are 117° and 8°, while for Fort Mohave, near the Colorado River (altitude 604 feet), they are 127° and 3°, respectively. The normal annual rainfall for Kingman is 11.50 inches, while for Fort Mohave it is 5.21 inches.

Juniper and scrub pine grow on the higher mountains and on favorable portions of the Plateau. Most of the county, however, is below 4,000 feet in altitude and supports only desert vegetation.

In places, Yucca or Joshua trees are abundant.

As shown by Figure 3, the Santa Fe Railway crosses Mohave County from east to west. Various highways, improved roads, and desert car trails traverse the area and lead to the mining districts (see Figure 3).

GENERAL GEOLOGY

The Plateau Province consists of essentially horizontal Paleozoic and Mesozoic strata, locally overlain by lava flows. In the Grand Canyon and Grand Wash cliffs, pre-Cambrian igneous and metamorphic rocks outcrop beneath these strata. The Grand Wash and Cottonwood cliffs, which limit the Plateau on the west, are fault-line scarps. The mountains west of these cliffs consist mainly of pre-Cambrian schist, gneiss, and coarse-grained granite, with local younger intrusive bodies, and Tertiary volcanic rocks. In the extreme southern portion of the county, some Paleozoic and Mesozoic metamorphic rocks are present. The intermont valleys or plains are floored with great thicknesses of loosely consolidated Tertiary and Quaternary sediments.

GOLD DEPOSITS

Mohave County ranks second among the gold-producing counties of Arizona. It has yielded nearly $40,000,000 worth of gold of which more than $37,000,000 worth has come from lode gold mines. The greater part of this production was made by the Oatman district.

As indicated by Figure 3, the gold districts are mainly in the west-central portion of the county. None occur within the Plateau Province.

Types: The gold deposits of Mohave County are of epithermal and mesothermal types.

The epithermal type prevails throughout most of the Black Mountains and is particularly well exemplified in the veins of the Oatman and Union Pass districts. The general characteristics of these veins are given on pages 83-89.

The mesothermal type prevails in the Cerbat Range (see pages 109-115).

LOST BASIN DISTRICT

The Lost Basin district is in a small northward-trending group of mountains, locally called the Lost Basin Range. This range rises east of Hualpai Wash and is separated from the Grand Wash

Cliffs by Grapevine Wash. Water for all purposes is hauled from Patterson Well or from the Colorado River, both of which are several miles distant.

The principal gold-bearing veins occur south of the middle of the belt in the western part of the area. They were discovered about 1886 and have been worked intermittently, but their production has been small. They strike northward, dip steeply, and occur in granitic and schistose rocks. The veins average from 4 to 6 feet in width, and some of them are traceable for more than a mile on the surface. Their croppings are principally iron and copper-stained quartz. The ore is reported to average about 0.4 ounces of gold per ton, with some silver.[98]

Schrader gives a brief description of the *Scanlon-Childers* prospect which, in 1915, was reported to have been taken over by the *Lost Basin Mining Company*.

GOLD BASIN DISTRICT

The Gold Basin district, in the eastern part of the White Hills, comprises a hilly area about 6 miles in diameter and from 2,900 to 5,000 feet in altitude. By road, it is some 60 miles from Kingman.

Since its discovery, in the early seventies, this district has produced about $100,000 worth of gold. The deposits are mesothermal veins in pre-Cambrian granite and schist. Their gangue is quartz, locally with siderite. The valuable metal is gold, mostly free milling, associated with limonite, malachite, cerussite, and, locally, vanadinite. Pyrite, chalcopyrite, galena, and molybdenite occur in places, but the water level has not been reached.

ELDORADO MINE[99]

The Eldorado mine is about 2 miles west of Hualpai Wash, at an altitude of 4,000 feet. It was discovered during the late seventies and was the first producer in the district. In 1907, it was owned by the Arizona-Minnesota Gold Mining Company. The total production is reported to have been $65,000 worth of bullion. Most of the ore was treated in the Basin or O. K. mill, in Hualpai Valley, 4 miles from the mine (see page 77).

At the mine, schistose, medium-grained, reddish granite is in fault contact with dark biotite granite. The principal vein strikes N. 50° E. and dips 65° SE. Developments on the vein include some 2,000 feet of adit tunnels and 40,000 cubic feet of stopes on three levels. The lowest stopes are approximately 200 feet below the highest point of the outcrop. As shown by these workings, the ore shoot averaged about 20 inches in width. It consisted of iron-stained gold-bearing quartz with malachite, cerussite, and vanadinite.

---

[98] Description abstracted from Schrader, work cited, pp. 150-51.
[99] Abstracted from Schrader, work cited, pp. 120-21.

O. K. MINE[100]

The O. K. mine, about ½ mile south of the Eldorado, was lo-cated in the early eighties. In 1886, a Kansas City company bought the property and built the O. K. mill in Hualpai Valley. This mill was operated intermittently until 1906, when it was de-stroyed by fire. Its ten stamps and cyanide plant were operated on water that was piped from springs in the Grand Wash Cliffs, 7 miles farther northeast. The O. K. Mine is reported to have produced $25,000 worth of gold.

The country rock is dark biotite granite. The vein strikes northeastward, dips about 75° NW., and averages about 18 inches in width. It is composed mainly of iron-stained quartz with cerus-site, siderite, galena, and molybdenite. The gold is commonly associated with cerussite.

Underground workings include about 1,600 feet of adit drifts, winzes, and stopes on four levels.

CYCLOPIC MINE

The Cyclopic mine is near the head of Cyclopic Wash, about 40 miles from Chloride. It was located during the eighties and has been intermittently worked by several concerns. In 1901, Robbins and Walker milled some of the ore. In 1904, the Cyclopic Gold Mining Company acquired the mine and later produced considerable bullion. During several years prior to 1921, inter-mittent production was made with a small cyanide mill. For some years after early 1923, the property was held by the Gold Basin Exploration Company. Intermittent production was made during 1932-1934.

The deposit occurs within a gently dipping brecciated zone in granite. This zone, as explored, extends to depths of 15 to 80 feet below the surface, and occurs discontinuously within an irregular northwestward-trending area about a mile long by 200 feet wide. In places, it is overlain by 5 to 15 feet of sand and gravel. The ore consists of brecciated fragments of coarse-tex-tured grayish vein quartz and country rock, more or less firmly cemented by iron oxide and silica. In places, it is cut by irregu-lar stringers of quartz. About 1,000 tons of ore that were recently mined are reported to have contained $4 in gold per ton. The gold is very fine grained.

Developments on the property include several open cuts and a 55-foot shaft; several old shafts from 40 to 50 feet deep; an old 300-foot incline that passed through the ore zone; and several hundred feet of old drifts and stopes, mostly within 30 feet of the surface.

OTHER PROPERTIES

The *Excelsior, Mascot, Never-get-left, Golden Rule, Gold Belt,*

---

[100] Abstracted from Schrader, work cited, pp. 121-22.

Senator, and *Salt Springs* properties, all of which have yielded small amounts of gold ore, are described by Schrader.[101]

NORTHERN BLACK MOUNTAINS

The Black Mountains continue for some 70 miles north of Union Pass, to Boulder Canyon. North of R. 25 N., they are limited on the west by the Colorado River, and are sometimes referred to as the River Range. This northern portion consists mainly of gneiss, schist, and granite along the eastern margin, overlain on the west by volcanic rocks and intruded by dikes of acid to basic composition.

Gold deposits occur in the Gold Bug, Mocking Bird, and Pilgrim localities, on the eastern side of the range, and in the Klondyke area, on the western side of the mountains. These deposits have received comparatively little geologic study.

The northern portion of the area is often termed the Weaver district, and the southeastern portion the Pilgrim district.

GOLD BUG MINE[102]

The Gold Bug mine is on the eastern slope of the Black Mountains, about 30 miles northwest of Chloride.

This deposit was discovered in 1893. Prior to 1895, it produced 50 tons of selected shipping ore that was worth $43,000. Later, about 800 tons of ore, treated in a 20-ton Huntington mill at the Colorado River, yielded returns worth $12,000. Some development work was done in 1908 and 1931. In 1932, a small tonnage of ore from the property was treated in the Kemple mill.

The deposit consists of two steeply northeastward-dipping veins, about 22 feet apart, in volcanic rock. Their gangue is mainly shattered quartz. The gold occurs principally near the hanging wall, associated with iron oxide in the oxidized zone and with pyrite and galena below the 200 level. It is generally accompanied by some silver. In places a little vanadanite occurs.

Prior to 1908, developments on the property included about 2,300 feet of underground work, with a 300-foot shaft and several shallower shafts.

MOCKING BIRD MINE

The Mocking Bird mine is 25 miles northwest of Chloride, in the foothills at the eastern base of the Black Mountains.

Prior to 1908, according to Schrader, this mine had made a reported production worth more than $20,000. He describes the deposit as follows:[103] "Its principal developments are twelve or fifteen shafts, ranging from 25 to 60 feet in depth, and about 500 feet of drifts. The vein lies nearly flat in a local sheet or flat-

---

[101] Schrader, F. C., work cited, pp. 121-27.
[102] Description largely abstracted from Schrader, U. S. Geol. Survey Bull. 397, pp. 217-18.
[103] Work cited, p 216.

lying dike of altered and pressed minette . . . Other volcanic rocks near-by consist of rhyolite tuff and latite. The vein is about 6 feet thick and consists of red and green quartz and breccia. The metal is gold with a small amount of silver. The gold occurs in a finely divided state, usually associated with hematite, of which much is present."

PILGRIM MINE

The Pilgrim mine, about 9 miles west of Chloride, at the eastern foot of the Black Mountains, was discovered in 1903. Prior to 1907, it was opened by a 360-foot inclined shaft, and a few tons of rich ore were produced. Some development work has been carried on intermittently during recent years, particularly by Pilgrim Mines, Inc., and by the Treasure Vault mining company. A small production was made in 1929. In 1933-1934, underground work was being done by the Pioneer Gold Mining Company.

The rocks of this vicinity consist of andesites and rhyolites, intruded by dikes of rhyolite-porphyry. The mineral deposit occurs within a fault zone that strikes N. 30° W. and dips 30° SW. In the foot wall and hanging wall of this zone are two veins, each of which range up to 3 or more feet in thickness. They are made up of irregular stringers and masses of fine-grained quartz with some calcite and silicified fragments of wall rock. The ore consists of epithermal fine-grained, greenish-yellow quartz. A small rich ore shoot was found on the 230-foot level, in the foot wall vein. There, the greenish-yellow quartz and fragments of silicified wall rock contained numerous small cavities filled with iron oxide and abundant visible gold.

About ½ mile southeast of the mine, the South Pilgrim Mining Company has prospected the vein with a 100-foot shaft and a short surface tunnel.

OTHER PROPERTIES

Schrader gives brief descriptions of the *Hall, Great West,* and *Pocahontas* mines, which are in the low foothills of pre-Cambrian gneiss, less than 2 miles south of the Mocking Bird mine.[104] Prior to 1908, each of these mines was developed to depths of about 200 feet. The Hall vein consists of a few inches to 2 feet of honey-combed quartz, locally rich in gold. In 1907, a 24-ton mill was in operation on the property.

At the *Golden Age* property, some development work was done in 1931 and 1932. Several years ago, a small production was reported from this mine.

At *Kemple Camp,* about 3½ miles southeast of the Mocking Bird mine, some rich ore has been mined and treated in a 5-stamp mill. The property is developed by several shafts, of which the deepest is 200 feet.[105]

Intermittent work has been carried on since 1931 at the *Mohave*

---

[104] Work cited, pp. 216-17.

Gold property, which is near the Gold Bug mine. Mr. Householder states that several small shipments of ore that contained
about 0.75 ounces of gold per ton, with some lead, have been
made.

The *Dixie Queen* mine, on the western slopes of the range,
yielded some shipping and milling ore several years ago. It also
made some production in 1927-1928 and 1931. During the past 2
years, it has been operated in a small way by lessees who treated
old tailings by amalgamation and cyanidation.[105]

The *Klondyke* mine, about 2 miles north of the Dixie Queen,
was operated about 1900. Mr. Householder states that about 4,500
tons of ore from the property were treated in the Klondyke
amalgamation mill, at the Colorado River, and later the tailings
were treated by cyanidation.

The *Golden Door* or *Red Gap* vein is in volcanic rocks, one mile
north of the Klondyke. It is said by Schrader[106] to consist of
more or less brecciated quartz, probably with adularia. In 1933,
lessees made a small production from this property.

OATMAN OR SAN FRANCISCO DISTRICT

SITUATION AND ACCESSIBILITY

The Oatman district, which includes the Vivian, Gold Road and
Boundary Cone localities, covers an area of about 10 miles long
by 7 miles wide on the western slopes of the southern portion of
the Black Mountains, in western Mohave County. It is also called
the San Francisco district, which is sometimes regarded as including also the Union Pass district, described on pages 101-108.

Oatman, the principal settlement, is 29 miles, via U. S. Highway 66, from Kingman, on the Santa Fe Railway. Numerous
secondary roads lead from this highway to the individual properties.

HISTORY AND PRODUCTION[107]

During the early sixties, soldiers from Camp Mohave, at the
Colorado River, carried on prospecting in this region. In 1863
or 1864, John Moss is reported to have taken $240,000 worth of
gold from a pocket in the Moss vein. The Hardy, Leland, and
Gold Dust veins were found soon afterwards, but the prominent

[105] Oral communication from E. Ross Householder.
[106] Work cited, p. 215.
[107] Largely abstracted from the following sources: Ransome, F. L., Geology of the Oatman gold district, Arizona: U. S. Geol. Survey Bull.
743, 1923.
    Lausen, Carl, Geology and ore deposits of the Oatman and Kather-
ine districts, Arizona: Univ. of Ariz., Ariz. Bureau of Mines Bull.
131, 1931.
    Schrader, F. C., Mineral deposits of the Cerbat Range, Black Moun-
tains, and Grand Wash Cliffs, Mohave County, Arizona: U. S. Geol.
Survey Bull. 397, 1909.
    Tenney, J. B., unpublished notes.

Year	Tom Reed		United Eastern		Gold Road	
	Tons Ore and Tails	Total Value Dollars	Tons ore	Total Value Dollars	Tons ore	Value Dollars
1897 to 1907					Approx. 100,000	$2,250,000
1908						
1909	Approx. 40,000	$ 1,037,911			Approx. 70,000	739,400
1910						
1911	43,924	835,048†			Approx. 120,000	676,600
1912	55,663	1,154,559†			109,070	665,783
1913	48,110	1,141,907†			103,629	676,515
1914	46,995	1,002,407†			107,846	843,991
1915	29,916	661,871†	Discovered		96,272	651,761
1916	46,170	486,678†	Developed		Developed	
1917	81,884	620,179†	84,543	$1,827,670	Developed	
1918	88,525	794,383†	92,339	2,072,359	Shut Down	
1919	89,557	679,986†	97,325	1,970,509		
1920	93,970	705,657†	102,926	2,233,819		
1921	69,832	377,992†	97,413	1,910,054		
1922	43,072	463,118	117,687	1,643,909	Mine Reopened	
1923	42,814	538,366	104,800	2,085,075	31,109	Approx. 150,000
1924	14,586	181,936	Closed June	Approx. 1,000,000	Closed	October
1925	35,448	494,829	Dump Ore Treated	Approx. 60,000		
1926	21,261	283,595	Leased	Approx. 50,000	Leased	
1927	17,259	161,461				
1928	7,672	118,275				
1929	Approx. 4,000	113,230				
1930	Approx. 20,000	Approx. 500,000				
1931	43,436	Approx. 700,000				
TOTAL	Approx. 984,090	Approx. $13,053,400	Approx. 697,038	Approx. $14,853,395	Approx. 737,926	Approx. $6,654,050

†Tom Reed production in fiscal year April to April.   *Union Pass District.

OATMAN DISTRICT.
by J. B. Tenney)

| | Total Production | | | |
Tons ore	Value Gold Dollars	Oz. Silver	Total Value Dollars	REMARKS Includes also production from the following:
100,000		Approx. 40,000	$ 2,522,000	Leland.
72,757	$ 266,254	6,522	269,711	Also Sheep Trail* and Victor.
18,106	300,036	7,118	303,737	Also Sheep Trail* and Victor.
89,284	1,103,221	26,254	1,117,398	Also Sheep Trail* and Victor.
110,699	1,458,639	33,834	1,476,571	Gold Crown and Ruth.
174,319	1,794,847	41,456	1,820,342	
159,984			1,818,522	
160,469			1,846,398	Gold Crown, Ruth, London.
132,579			1,499,033	Frisco*, Banner, Ruth.
95,245	892,681	23,812	908,349	Gold Dust, Orphan*.
167,258	2,310,270	57,353	2,357,529	
182,824	2,772,991	70,432	2,843,423	Gold Ore, New Philadelphia, Orphan,* Pioneer.
184,490	2,556,197	71,883	2,636,650	Arabian,* Gold Trails, Pioneer.
197,629	2,830,731	92,806	2,931,890	Green Quartz, Thumb Butte.
179,013			2,388,050	
169,240			2,138,546	United American, Telluride, Oatman United.
186,686	2,796,830	68,551	2,853,042	Oatman United, Gold Dust, Orphan.*
96,783	1,617,196	39,097	1,643,391	Telluride, United American, Gold Dust.
33,826	502,019	11,721	510,153	Lexington, Apex, Pioneer.
29,721	395,971	9,964	402,188	Oatman United, Gold Dust, Sheep Trail.*
15,028	147,599	4,708	150,268	
11,817	147,389	4,152	149,818	Western Apex, United American.
4,430	118,516	4,068	120,684	Sunnyside, Vivian, Arabian,* Keystone.
28,048	580,768	18,274	587,803	United American, Pioneer, Telluride, Vivian.
45,414	706,767	21,771	713,106	Pioneer, United American, Big Jim, Sunnyside.
2,645,613	Approx. $35,740,000		Approx. $36,008,602	

PLATE I

GEOLOGIC MAP OF THE
OATMAN DISTRICT
MOHAVE COUNTY, ARIZONA.

LEGEND

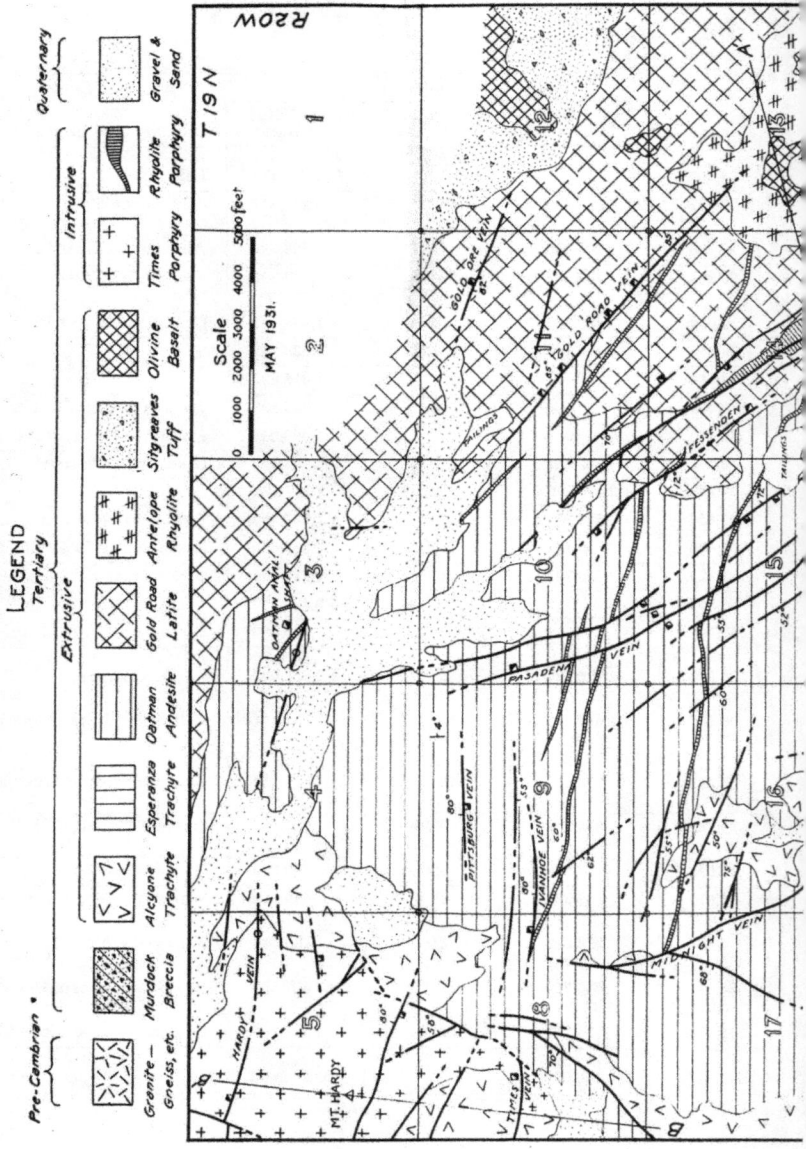

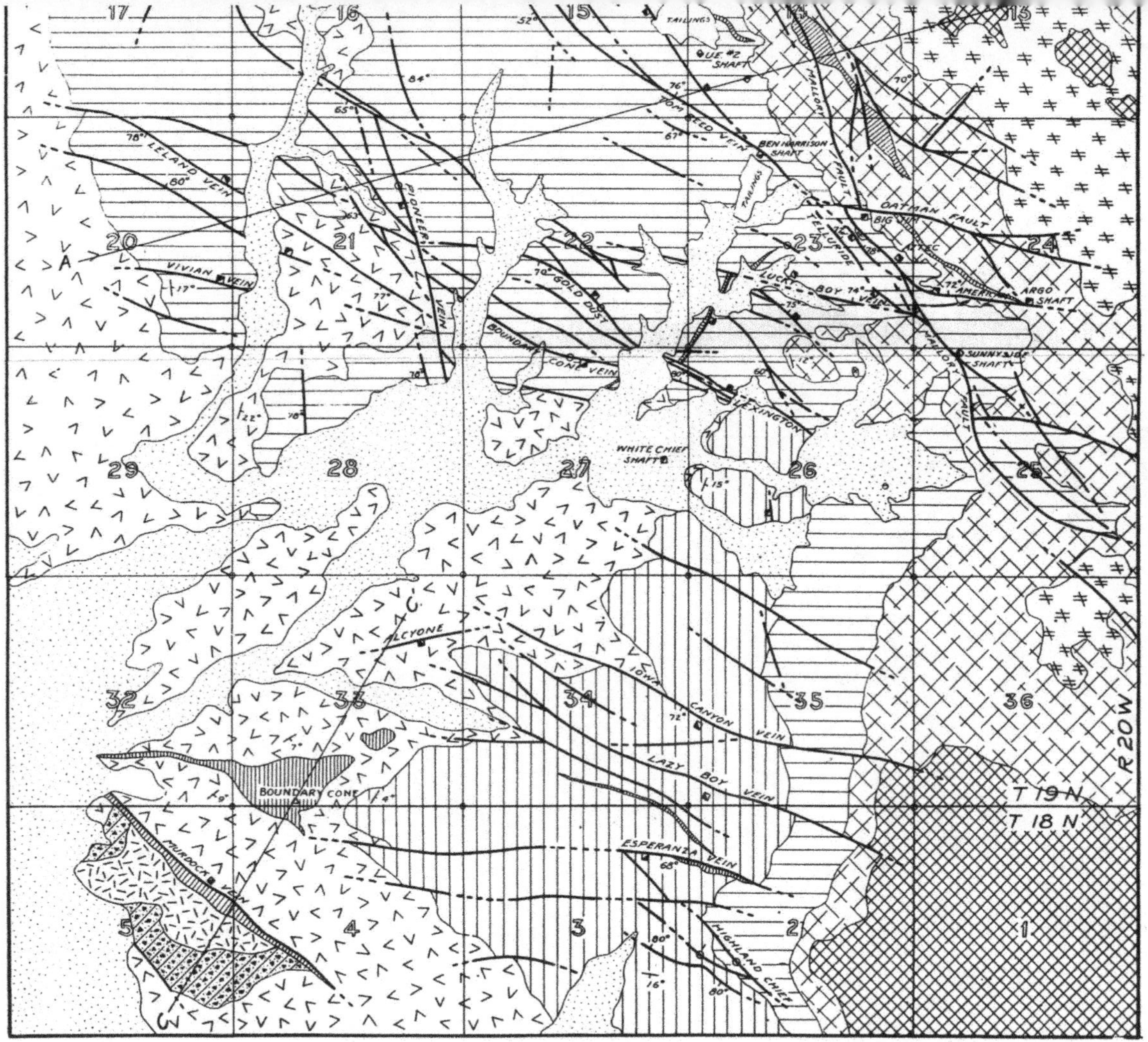

MALLORY FAULT
OATMAN FAULT
TOM REED VEIN
BEN HARRISON SHAFT
UE #2 SHAFT
TAILINGS
TELLURIDE
BIG JIM
AZTEC
AMERICAN
ARGO SHAFT
SUNNYSIDE SHAFT
LUCKY BOY VEIN
GOLD DUST
BOUNDARY CONE VEIN
LEXINGTON
WHITE CHIEF SHAFT
PIONEER VEIN
LELAND VEIN
VIVIAN VEIN
ALCYONE
IOWA CANYON VEIN
LAZY BOY VEIN
BOUNDARY CONE
ESPERANZA VEIN
HIGHLAND CHIEF
MURDOCK VEIN
R 20 W
T 19 N
T 18 N

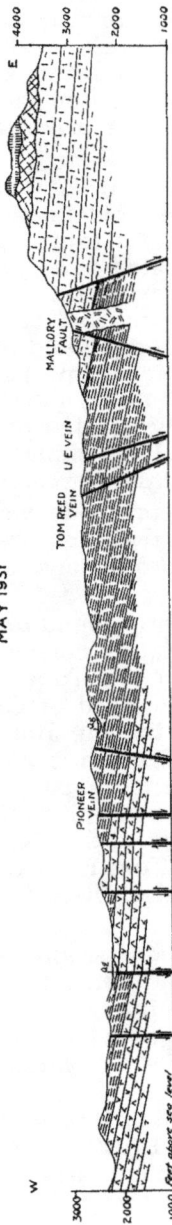

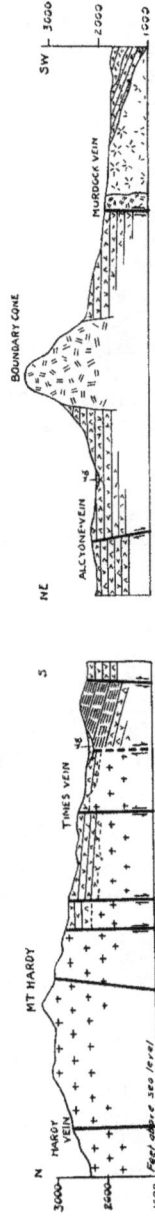

Figure 4.—Structure sections of the Oatman district, along lines indicated on Plate I, by Carl Lausen

outcrops of the Tom Reed and Gold Road veins remained un-
tested for many years. The town of Silver City grew up at a
watering place on Silver Creek, about one mile south of the Moss
lode, and a small mill was established at Hardyville, on the Colorado River. Some ore was treated in arrastres and in this mill,
but the results were disappointing.[108] After the 1866 outbreak
of the Hualpai Indians, the district was practically abandoned
for several years.

A revival in activity took place in 1900 when rich ore was found
in the Gold Road vein. In 1901, the Gold Road Company sank
the Tom Reed and Ben Harrison shafts to a depth of 100 feet.
The Leonora mill, at Hardyville, operated during part of 1901 and
1902 on ore from the Moss and Hardy veins. During 1903 and
1904, the Mohave Gold Mining Company did considerable work
on the Leland property. The Blue Ridge Gold Mines Company
produced ore from the Tom Reed vein during part of 1904-1905.
In 1906, the Tom Reed Gold Mines Company purchased the mine,
developed high-grade ore, and, in 1908, started production which
continued through 1931. The Gold Road mine produced intermittently until 1916. The town of Oatman was started about 1912.

During 1915 and 1916, a $6,000,000 ore body was developed in
the United Eastern mine. The fact that this ore shoot did not
outcrop prompted scores of wildcat promotions, but these efforts
proved to be largely futile.

In 1916, the Big Jim Mining Company found an important ore
body on their Big Jim claim, immediately northeast of the Grey
Eagle and the Black Eagle claims of the Tom Reed Company.
Further work indicated that the Tom Reed or Grey Eagle vein
is the upper, downfaulted portion of the Big Jim vein. The displacement, principally along the Mallery fault, amounts to about
400 feet. In 1917, the United Eastern Company purchased the
Big Jim ground, but two years later the Tom Reed Company
brought suit to establish its apex claim to the Big Jim vein. The
courts, however, decided against the Tom Reed Company on the
grounds that the amount of horizontal displacement could not be
proven.

In 1924, the United Eastern ore body became exhausted, and
the mine was closed. Considerable diamond-drill prospecting
was done, but with unsatisfactory results.

During the past several years, except for an interval from early
1932 to early 1934, the Tom Reed 250-ton cyanide mill has run
partly as a customs plant.

During 1933, and early 1934, the principal operations in the district were carried on by lessees in the Big Jim mine, the ore from
which was treated in the Telluride mill. Early in 1934, development was being carried on at several properties, principally the

---

[108] Raymond, R. W., Statistics of mines and mining in the states and
territories west of the Rocky Mountains, 1871, p. 265.

Tom Reed and Rainbow mines.

As shown in the accompanying table, the Oatman district to the end of 1931 produced more than $34,500,000 worth of gold The 1932 production amounted to $71,410.

TOPOGRAPHY AND GEOLOGY[109]

The southern portion of the Black Mountains consists of a very ruggedly dissected, gently eastward-dipping block of Tertiary volcanic rocks which rest upon a basement of pre-Cambrian gneiss and granite.

The Oatman district is in a belt of rugged foothills at the western base of the mountains, mainly between altitudes of 2,000 and 3,200 feet. Eastward, the range rises with deeply dissected, step-like cliffs to a maximum elevation of about 5,000 feet above sea level or 4,500 feet above the Colorado River.

Gulches which are dry except during rainy seasons carry the run off from this side of the mountains westward to the Colorado River. A few small perennial springs issue from tuffaceous beds, particularly in Silver Creek.

The principal formations, as mapped by Ransome and by Lausen, consist of a few patches of gneiss and granite on the west, overlain by a thick series of trachytes, andesite, latite, tuffs, rhyolite, and basalt. Intrusive into parts of this series are monzonitic, granitic, and rhyolitic porphyrys. The relations of these rocks are shown by Lausen's map (Plate I) and sections (Figure 4). The most important ore-bearing formation is the Oatman andesite which Schrader[110] termed the "green chloritic andesite."

These formations are cut by numerous faults of prevailingly northwestward strike and steep northeastward dip.

DISTRIBUTION OF VEINS

Ransome states that the veins of the Oatman district occur within fissures along which faulting has taken place, as a rule before, during, and after the period of vein formation. In general, no sharp distinction between faults and veins can be made, although some fissures, such as the Mallery fault, are younger than the veins. As indicated on Plate I, the veins occur rather widely distributed, and the most productive ones are in the northeastern half of the district.

FORM OF VEINS

Some of the veins have tabular forms but the larger ones are essentially stringer lodes of complex structure. Compound veins, which consist of two or more veins separated by country rock with stringer veinlets, are common (see Plate II). Many of the veins are lenticular in all dimesions; a strong vein may pinch out within a few tens of feet, and an insignificant stringer may

---

[109] Largely abstracted from Ransome, F. L., work cited, pp. 8-32; Lausen, Carl, work cited, pp. 18-55.

[110] Schrader, F. C., work cited, pp. 34-37.

thicken to considerable width within a distance of 30 feet. It is stated that few of the veins attain maximum widths of more than 50 feet. Some of the outcrops, as that of the Gold Road vein, are very conspicuous, but others, like that of the United Eastern, are scarcely noticeable.

Only very small, commercially unimportant placers have been formed.

MINERALOGY

The gangue of the Oatman vein consists mainly of quartz and calcite, either of which may predominate. According to Ransome, the vein material that consists entirely of quartz or calcite is generally of very low grade or barren. Microscopic adularia is a common constituent of the gold-bearing quartz. Fluorite occurs in some of the veins, but is very rare in the larger ore bodies. Gypsum and kaolin are locally abundant in the oxidized zone.

The metallic minerals consist of free gold and rare pyrite and chalcopyrite. This gold is characteristically fine grained and generally can be seen only in rich ore.

STAGES OF MINERAL DEPOSITION

Ransome[111] says: "The individual veins and stringers appear banded in cross section, showing that the vein minerals were deposited in successive layers from the walls to the middle of the fissure.

"In a broad way in the Oatman veins the deposition of fine-grained white quartz, which has, in part at least, replaced andesite and contains little or no gold, has been followed by the deposition of the gold-bearing quartz accompanied by some calcite and adularia, followed in turn by barren calcite. This general sequence, however, is certainly far from being the complete record."

Lausen,[112] after further detailed work in the district, has recognized five stages of vein filling, each of which has its distinctive type of quartz. He summarizes their characteristics in the following table:

---

[111] Work cited, pp. 33-34.
[112] Work cited, pp. 63-72.

SUMMARY OF THE CHARACTERISTIC FEATURES OF THE VARIOUS STAGES OF QUARTZ DEPOSITION.

Stage	Texture	Color	Range of oz. gold per ton.	Ratio of gold to silver.	Relative Distribution in the veins.
1st	Coarse to fine grained.	Colorless, white amethystine.	Up to 0.06	1 to 6	Abundant.
2nd	Fine grained. Often shows casts of calcite.	White, rarely yellow	Up to 0.08	1 to 6	Abundant.
3rd	Fine grained Banded	Various colors.	0.06 to 0.40	2 to 3	Relatively scarce.
4th	Fine grained. Often shows casts of platy calcite.	Pale green to yellow.	0.20 to 1.00	1 to 2	Abundant only in ore shoots.
5th	Fine to medium grained. Usually banded.	Pale to deep honey-yellow.	1.00 up	4 to 1	Abundant only in ore shoots.

Detailed descriptions and photographs of these five stages of quartz are given in Lausen's report. Only the commercially important third, fourth, and fifth stages will be considered here.

Plate II.—View showing structure of the Gold Road vein, Oatman district.
Photograph by Carl Lausen.

Third stage: Lausen says: "This variety of quartz is extremely fine grained, often chalcedonic, and consists of alternate layers of slightly different color. (See Plate III). Usually, it is a creamy white with thin bands of pale brown. Sometimes the broader bands are a delicate lavender between layers of white and yellow . . . Very thin layers or partings of calcite may be seen in some specimens . . . This type of quartz has a rather limited distribution. In the Gold Road and Gold Ore veins, much of the vein filling is of this type. Smaller amounts may be seen at various other mines, notably at the Pioneer.

"Values of this type of quartz range from 0.06 to 0.40 ounces in gold and 0.18 to 0.24 ounces of silver per ton. The average ratio of gold to silver is 2 to 3.

"Calcite again followed the deposition of quartz and is usually flesh-colored. It occurs as thin plates, sometimes as much as 6 inches across."

Plate III.—Banded quartz of the third stage of deposition, Oatman district.
Photograph by Carl Lausen

Fourth stage: Lausen describes this quartz, which contains pseudomorphs of platy calcite (see Plate IV), as follows: "The

Plate IV.—Quartz of the fourth stage of deposition, Oatman district.
Photograph by Carl Lausen

color is invariably yellow or greenish, but the oily luster is absent except in such specimens as have a banded structure. The banded structure is best seen where the earlier stages of mineral filling have been shattered and this later quartz introduced into the fractures.

"A microscopic examination of this quartz shows . . . occasional grains and crystals of adularia.

"In the Oatman district, this type of quartz occurs at practically all the mines that have produced gold . . . Assays of such quartz range from 0.20 to 1.00 ounces in gold and 0.24 to 2.34 ounces in silver per ton. The ratio of gold to silver is approximately 1 to 2.

"The calcite formed upon this pseudomorphic quartz occurs as very thin lamellae which form a somewhat compact mass of curved plates. The calcite has a pearly luster and is the most distinctive variety of this mineral in all the stages of mineral deposition in these veins. A broad band of the calcite, several inches wide, may be traversed by thin layers of the yellow quartz which show a rhythmic alteration of quartz and calcite."

"*Fifth stage:* The vein filling of this stage of mineralization was deposited in open fissures upon the earlier stages of vein filling or upon rock fragments. Banding is well developed and crenulation in the bands is very pronounced.

Plate V.—Quartz (dark) and adularia (white band) of the fifth stage of deposition, Oatman district.

Photograph by Carl Lausen.

"The quartz of this last stage of mineralization is yellow and, often, olive-green in color. It invariably shows an oily luster . . . The size of the quartz grains varies from fine to coarse, and, usually, the darker-colored bands are somewhat coarser in texture . . . Layers of quartz are often separated by bands of snow-white adularia that vary in width from a small fraction of an inch up to 2 inches. Occasionally, the quartz and adularia are separated by a thin parting of calcite.

"Gold is often concentrated in certain bands as clusters of small grains. Adjoining bands of quartz may contain only isolated grains . . . Usually the coarse crystals of adularia contain very little gold.

"Assays of this stage of deposition always show commercial values and range from one ounce of gold per ton up . . . The average ratio of gold to silver is 4 to 1.

"The calcite that followed the deposition of the fifth-stage quartz is transparent, colorless, and usually well crystallized."

WALL-ROCK ALTERATION

In general, the wall rocks of the veins show alteration to quartz, chlorite, and pyrite. Ransome mentions also sericite as a wall-rock alteration product in the Oatman andesite.

ORE SHOOTS

The ore shoots of the Oatman district are characteristically lenticular in plan and irregular in longitudinal section. Many of them are narrow but some are wide, as, for example, the United Eastern ore body which had a maximum width of 48 feet. The most productive ore shoots of the district were in the United Eastern, Tom Reed, and Gold Road mines. Their distribution and approximate size in these mines, as indicated by areas of stoping, are shown on Figure 5. Ransome states that good ore has been found in some ten or twelve veins, and some are credited with outputs amounting to a few thousand dollars, but these ore bodies have been small.

As recounted in the descriptions of the individual mines, the gold and silver content of the ores decreased rather sharply with depth. Very little ore has been found in the district below a depth of 1,000 feet. The average value of the ore mined from 1908 to 1928 was $12.37 per ton.

Although the largest known ore bodies occur in the Oatman andesite, productive ones have been found in various other formations. Vein-intersections appear to have been generally barren. The ore shoots were evidently localized within certain relatively permeable channels which resulted from faulting, but an understanding of the various structural features involved will be possible only after much further detailed geologic work.

ORIGIN OF THE ORES

The veins of the Oatman district are of the epithermal bonanza type characteristic of Tertiary volcanic activity. The ores were deposited by ascending thermal solutions at depths of not more than about 3,000 feet below what was then the surface. The interpretation of the rather limited vertical range of deposition is that of rapid decrease in the temperature and pressure of the solutions near the surface.

Ransome found no evidence of downward enrichment of the gold. In fact, the general abundance of calcite in these veins would tend to prevent any important supergene enrichment of the gold except where there existed channels that were inert to acid solutions. Furthermore, the amount of pyrite available to form acid is very small in comparison to the amount of calcite in the veins.

MINES OF THE OATMAN DISTRICT[113]

TOM REED PROPERTY

History and production: The first work done on the Tom Reed vein was in 1901 when the Gold Road Company sank the Tom Reed and Ben Harrison shafts to a depth of 100 feet. The Blue Ridge Gold Mines Company bonded the property in 1904, built a 20-stamp mill, and did some developing and stoping. The Tom Reed Gold Mines Company purchased the property in 1906 and started producing in 1908. Production continued rather steadily until March, 1932, since when operations have consisted largely of development work. Early in 1934, development was being carried on in the Grey Eagle and Black Eagle ground, and the mill was treating customs ores. As given in the table on page 82, the total yield of the Tom Reed property from 1908 through 1931 amounted to approximately $13,053,000.

Ore bodies: As shown on Plate I, the Tom Reed vein strikes northwestward and dips steeply northeastward. Except for a small section on the Black Eagle claim, the vein, so far as explored, occurs in Oatman andesite.

On the Pasadena claim, at the northwestern end of the vein, ore was found to a depth of only 55 feet.

Exploration of the northwestern portion of the vein on the 300- and 500-foot levels of the Olla Oatman shaft failed to find any ore body.

The Tip Top ore body, beneath the town of Oatman, extended to below the 1,400-foot level where it ended in a rather sharp point (see Figure 5).

The Ben Harrison ore shoot extended with a rather irregular outline to a depth of 800 feet and was 16 to 20 feet wide in the shaft. It averaged $25 per ton for the first 30 feet of depth and $12 per ton below that level.

On the Aztec Center claim, the Big-Jim-Aztec ore shoot was 900 feet long by 10 to 20 feet wide and extended from the Mallery fault, near the 300-foot level, to below the 600-foot level. Much of the ore belonged to the fourth and fifth stages of deposition and may have averaged more than $12 per ton. On the Grey Eagle-Bald Eagle claims the supposedly down-faulted segment of this ore shoot averaged more than $20 per ton. The displacement along the steep northwestward dip of the Mallery fault amounts to about 500 feet, but the horizontal displacement is unknown.

The Black Eagle ore body, south of the Aztec shaft, has been developed to the 1,100-foot level. On the 900-foot level, it is

---

[113] The following information is largely abstracted from Lausen, Ransome, and Schrader, works cited, with added facts on recent operations. Most of the data on history and production are taken from unpublished notes of J. B. Tenney.

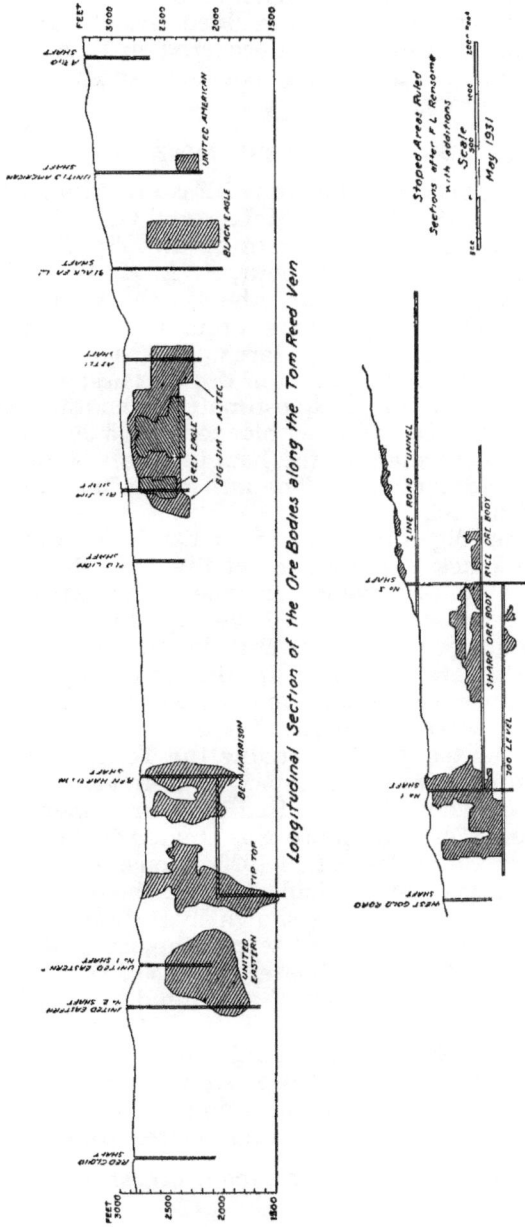

Figure 5.—Longitudinal sections showing location of ore shoots in the Tom Reed vein and the Gold Road vein, Oatman district, by Carl Lausen, 1931.

more than 400 feet long by an average of 7 feet wide. Much of the ore was of high grade, with visible free gold.

South of the junction of the Tom Reed and Telluride veins, a part of the Telluride ore shoot that occurred on Tom Reed ground extended from the 200- to the 500-foot level and was only 3 feet wide but of good grade.

UNITED EASTERN MINE

History and production: The United Eastern Mining Company was incorporated in 1913 by J. L. McIver and G. W. Long to prospect an apparent fault fissure north of the Tip Top ore body. Aided financially by W. K. Ridenour, F. A. Keith, S. W. Mudd, P. Wiseman, C. H. Palmer, Jr., and others, a shaft was sunk at the northern end of Oatman and the ground was purchased for $50,000. Ransome says:[114] "In March, 1915, a cross-cut on the 465-foot level went through 25 feet of ore that assayed $22.93. By the end of 1916, after an ore body estimated to contain gold to the value of about $6,000,000 had been blocked out, a 200-ton mill had been built and a new shaft in the hanging wall of the vein had been completed and equipped. The mill was afterward enlarged to a capacity of 300 tons."

Meanwhile, the Big Jim vein, immediately northeast of the Grey Eagle and Black Eagle claims of the Tom Reed Company, had been discovered. Ransome continues: "In April, 1917, the United Eastern Company purchased the Big Jim mine. The able mining engineers who were directing the United Eastern Company were well aware that the Big Jim and Grey Eagle veins were probably once continuous and owed their separation to faulting, but they were convinced that the amount of the throw about 400 feet, was too great to enable the Tom Reed Company to make successful claim to the Apex of their vein."

The known ore bodies of the United Eastern property were exhausted by June, 1924. Dump ore was treated in 1925, and some production was made by lessees in 1926. Considerable diamond-drill prospecting was done, particularly from the eighth and tenth levels, but without success. As shown in the table on page 82, the total yield of the United Eastern mine from 1917 to 1926, inclusive, amounted to approximately $14,853,395. The average costs per ton, according to Moore,[115] amounted to $8.254 of which $4.332 was for mining.

Since late 1932, lessees, employing about thirty men, have mined approximately 50 tons of ore per day from between the 500- and 700-foot levels of the Big Jim mine. This ore was milled in the Telluride plant of the Oatman Associates Mining Company.

Ore bodies: The United Eastern main ore shoot occurred on

---

[114] Work cited, p. 6.
[115] Moore, R. W., Mining methods and records at the United Eastern mine: A. I. M. E., Trans., vol. 76, pp. 56-90. 1928.

the Tom Reed Extension claim, on the northeastern branch of the Tom Reed fracture, along which the Gold Road latite has been dropped against the Oatman andesite. In maximum dimensions, this ore shoot was 950 feet long, 750 feet high, and 48 feet wide. It nowhere outcropped at the surface as a distinct vein and was of commercial grade only between the third and ninth levels. The vein continues in depth, but only as low-grade material. The ore consisted partly of massive quartz with adularia and calcite and partly of stringers separated by barren andesite. This ore body produced 511,976 tons with an average gross value of $21.037 per ton.[116]

The other ore body mined by this company was on the Big Jim claim. As already explained, the upper portion of this shoot had been displaced by faulting and was mined on the Grey Eagle and Bald Eagle claims. In maximum dimensions, the ore body on the Big Jim claim was 850 feet long, 450 feet high, and 35 feet wide. It produced 220,552 tons with an average gross value of $17.248 per ton.

GOLD ROAD MINE

History and production: The Gold Road property consists of 4 claims on the Gold Road vein, in the northeastern part of the district and about 1½ miles northeast of the Tom Reed.

Rich gold ore was discovered in this vein during 1900. The ground was worked in a small way for several months by various people, and, in 1902, the Gold Road Mining and Explorations Company was incorporated to carry on the operations. In 1911, the property was purchased by the Needles Smelting and Refining Company, a subsidiary of the United States Smelting, Refining, and Mining Company. By the end of 1916, the known ore bodies were nearly exhausted, and the mine was closed. It was again worked during 1922 and 1923. Since 1923, only intermittent operations by lessees have been carried on.

As shown in the table on page 82, the Gold Road property to the end of 1923 produced $6,654,050 worth of bullion.

Ore bodies: The Gold Road vein strikes about N. 50° W., dips 80° to 85° NE., and occupies a fault zone in latite. The vein is a stringer lode with a prominent outcrop, in places nearly 100 feet wide. Most of the vein filling occurs in two zones, termed the north and south veins, which are separated by largely barren latite with minor stringers of quartz.

The vein filling, as seen in the outcrops, is largely chalcedonic, wavy banded, vari-colored quartz of the third stage of deposition. In the ore shoots, quartz of the fourth and fifth stages was present. Calcite occurs in distinct veinlets younger than the quartz.

Most of the ore mined came from three shoots on the north vein. The largest of these shoots was in the vicinity of No. 1

---

[116] Moore, R. W., work cited, p. 71.

shaft, as shown in Figure 5. This shoot was 900 feet long by a maximum of 22 feet wide and extended from the surface to the 700 level. Some 600 feet farther southeast is the Sharp ore body which, as shown on Figure 5, was mined from the 300 to the 500 level and from the 700 to the 800 level. Less than 200 feet farther southeast, between the 300 and 500 levels, was the Rice ore body which was about 400 feet long. For about 1,500 feet farther southeast, small pockets of ore were mined from near the surface. As indicated by Figure 5, the Line Road tunnel prospected the vein below these outcrops for a horizontal distance of 3,000 feet, but found no commercial ore.

MOSS MINE

The Moss mine is about 7 miles northwest of Oatman and 2 miles north of Silver Creek.

This deposit was probably the first to be worked in the district. During the early days, it made a reported production of $240,000 from near the surface. Since that time, considerable intermittent development has been carried on, but little ore has been mined.

The vein strikes N. 78° W., dips 70° S., and occurs in the Moss quartz monzonite-porphyry. It forms a lode from 20 to more than 100 feet wide, with the widest portion at the western end, and is traceable on the surface for more than a mile east of the mine.

The vein filling consists of fine-grained white quartz and calcite, with stringers of colorless to pale-green fluorite. The largest ore shoot consisted of free gold in iron-stained quartz but extended to a depth of only 65 feet. Several smaller ore bodies were mined from near the surface at various places along the vein.

Workings on the Moss vein include a 230-foot shaft with about 750 feet of workings, about 900 feet of tunnel, and some irregular surface openings. Ransome states that the vein on the 220 level appears to be 90 feet wide and as a whole probably carries from 0.15 to 0.20 ounces of gold per ton.

TELLURIDE MINE

The Telluride vein, which joins the Lucky Boy and Tom Reed veins south of the Ben Harrison shaft, has made an estimated gross production of about $200,000. The mine was active from 1922 to 1925, and was operated in a small way during 1930-1934.

This vein strikes northwestward with an inconspicuous outcrop, and is about 3,000 feet long. The ore, which occurred mainly between the 300- and 500-foot levels, ranged from a few inches to 5 feet in width. The vein filling consists mainly of quartz and calcite of the first and second stages of deposition, accompanied by a 6-inch streak of fifth-stage quartz.

PIONEER OR GERMAN-AMERICAN MINE

The Pioneer mine, formerly known as the German-American, is about 1½ miles southwest of Oatman.

In 1902, the Gold Road Company did some development work

on this property. During 1903-1906, the German-American Mining Company produced 2,700 tons of ore that averaged about $10 in gold per ton. Some production was made by lessees in 1925, 1930, and 1932.

The Pioneer vein strikes N. 13° W., and dips 80° E. Its hanging wall is Oatman andesite and its foot wall for 2,000 feet is Alcyone trachyte. The vein intersects the Gold Dust-Boundary Cone fault zone at an angle of about 40°, but neither vein shows offsetting by the other. Near the Pioneer shaft, the vein has its maximum width of about 18 feet. There, it consists of coarse-grained, gray calcite and quartz that are of rather low grade except in iron-stained zones of crushing.

Several small ore bodies were mined towards the southern end of the Pioneer vein. One of them, near the Treadwell shaft, was about 400 feet long by a maximum of 3½ feet wide, but terminated abruptly near the 400 level. Specimens of this ore consist of quartz and unreplaced calcite, with some fourth-stage, greenish quartz.

North of the Thirty-fifth Parallel shaft, a shoot about 200 feet long by 3 feet wide yielded iron-stained ore from near the surface.

Near the Pioneer shaft, at the northern end of the property, narrow portions of the vein assay more than $10 per ton and yielded some relatively rich ore from near the surface.

GOLD DUST MINE

The Gold Dust mine, formerly known as the Victor-Virgin and the Orion, is about a mile southwest of the Ben Harrison shaft.

This deposit, which was located in the early days, produced a small amount of shipping ore prior to 1907. According to local press reports, development was carried on intermittently from 1916 to 1924. Some production was made in 1923-1926 and 1932, but the total output is unknown. Early in 1934, part of the mine was being reconditioned by lessees.

Lausen[117] gives the following description: "The general trend of the vein is northwest, and, on the Virgin Lode claim, the vein branches. The southern branch, striking nearly east-west, intersects the Pioneer vein a few hundred feet south of the Pioneer shaft. The north branch continues northwest and appears to be continuous with the southern extremity of the Midnight vein. The country rock throughout the entire length of this vein is the Oatman andesite.

At some places, the vein consists of solid quartz and calcite with a width up to 7 feet. To the northwest, the vein splits up into a series of small stringers which ramify through a zone of somewhat altered andesite over 20 feet wide. According to Ransome[118] two ore shoots were mined, the largest having a length of 200 feet and extending from the surface to a depth of 160 feet

---

[117] Work cited, pp. 110-11.
[118] Work cited, p. 49.

A small ore shoot, mined near the No. 2 shaft, extended only from the surface to the 100-foot level. This ore shoot occurs up the hill to the northwest of the main shaft and at a higher elevation. The vertical range of the ore shoots as mined is approximately 500 feet. Development work from the 500-foot level of the main shaft failed to find ore bodies in depth on this vein. The ore mined consisted of greenish quartz with unreplaced remnants of calcite. Adularia is present in the specimens collected only as a microscopic constituent."

LELAND MINE

The Leland property, also known as the Leland-Mitchell, is about 2 miles west of Oatman, as shown on Plate I.

In 1871, Raymond[119] stated that considerable unsuccessful work had been done on the Leland and Mitchell lodes. The property produced some high-grade ore in 1902. In 1903, the Mount Mohave Gold Mining Company, afterward the Mohave Gold Mining Company, obtained the property, erected a mill at Milltown, 11 miles west of the mine, and built 17 miles of narrow-gauge railway to the Colorado River. The venture proved unsuccessful, however, and the mine was closed about the end of 1904. A production of about $40,000 is reported to have been made from 4,500 tons of ore.

The Leland vein strikes about 30° SE. and dips 70° SW. The ore that was mined came from stopes worked through tunnels in Leland Hill. There, the vein is a stringer lode, up to 15 feet wide, without definite walls. The vein material consists of quartz and calcite. Some of this quartz is of greenish color.

The Mitchell vein, which outcrops about 500 feet south of the Leland, strikes southeastward, dips northeastward, and has a maximum width of 7 feet. It was opened by a shaft more than 700 feet deep, but produced only a little ore from near the surface. The 700 level is said to be under water.

MIDNIGHT MINE

The Midnight mine[120] is about 2½ miles northwest of Oatman. In 1907, it was the property of the Mohave Gold Mining Company and had been producing moderately since about 1900. It has long been idle.

The vein, which is the foot wall member of a wide lode, strikes N. 15° W., dips 30° to 40° W., and occurs in Oatman andesite and probably also in Alcyone trachyte. The filling consists of quartz, calcite, and fluorite. Near the foot wall, the quartz is dark and chalcedonic. It was stoped from near the surface to a depth of approximately 50 feet by a length of 120 feet and a width of 3½ feet. This ore is stated to have averaged about $18 per ton in

---

[119] Raymond, R. W., Statistics of mines and mining in the states and territories west of the Rocky Mountains, 1871, p. 265.

[120] Abstracted from Schrader, work cited, p. 194.

gold, but material from a depth of about 50 feet ran only $7.

SUNNYSIDE MINE[121]

"The Sunnyside Mine is in the southeastern part of the district, near the center of Section 25. The fault on which this mine is located has a known length of about one mile. At the north end, it is offset 100 feet by an east-west fault, and beyond this fault, it is difficult to trace. At the south end, it joins the Mallery fault. The vein is well exposed at several places along the fault and appears to be continuous, but it is difficult to trace. The wall rock at the surface is andesite.

"Much of the vein consists of coarse-grained quartz and calcite and is low grade. The vein is quite variable in width along the strike, and, at one place, it contains 5 feet of solid quartz and calcite. Small stringers occur in the adjacent wall rock.

"The main shaft was sunk to a depth of somewhat over 500 feet. On the 500-foot level, the footwall of the vein, which at this point dips steeply to the northeast, is the Esperanza trachyte, while the hanging wall is Oatman andesite.

"In the spring of 1928, a small ore shoot was found in a winze sunk from the 500-foot level. The ore shoot had a rather limited extent, both horizontally and vertically. The ore consisted of quartz and very little calcite, and specimens showed visible specks of coarse gold. To the east of this ore shoot, is a lode 40 feet wide, which consists of quartz stringers in andesite. The ore is too low in grade to be mined. The small ore shoot was later mined by lessees." Some production was made in 1929 and 1930.

IOWA MINE[122]

The Iowa property is on the Iowa Canyon vein, in the southern part of the area mapped in Plate I. Prior to 1907, it was developed by a 200-foot shaft with about 100 feet of drifts. A little development work was done in 1916.

The vein strikes west-northwestward, dips about 70° S., and occurs in Esperanza trachyte.

According to Schrader, the filling ranges from 3 to 8 feet in width and consists of greenish brecciated quartz, a little calcite, and some rock breccia, traversed by veinlets of quartz and calcite. He states that its 3 feet of width adjacent to the hanging wall contains from 0.15 to 0.70 ounces of gold per ton.

LAZY BOY MINE[123]

The Lazy Boy vein is about ¼ mile south of and parallel to, the Iowa. It is reported to be opened by a 200-foot shaft with some drifts. Some development work was done in 1916. Schrad-

---

[121] Quoted from Lausen, work cited, p. 112.
[122] Largely abstracted from Schrader, work cited, pp. 200-201.
[123] Largely abstracted from Schrader, work cited, p. 201.

er states that the vein filling in general is similar to that of the
Iowa and in places is reported to contain nearly 0.5 ounces of
gold per ton.

HARDY VEIN[124]

The Hardy vein outcrops in the northwestern portion of the
area shown by Plate I. It is reported to have yielded rich ore
from near the surface during the early days. Prior to 1907, it was
explored to a depth of about 300 feet by several shafts with more
than 1,200 feet of drifts, and by several hundred feet of adit
tunnels.

The vein strikes west-northwestward and dips 75° N. It occurs
in the Times micrographic granite-porphyry and ranges from 2
to 30 feet wide. Its filling consists of quartz, calcite, and
fluorite. Locally, adularia is present. Where prospected, the
vein is reported to contain from 0.3 to 1.5 ounces of gold per ton.

GADDIS-PERRY VEIN[125]

The Gaddis-Perry vein, between Silver Creek and the Hardy
vein, is a short distance beyond the northwestern limit of the area
included in Plate I.

Here, the prevailing rock is the Times micrographic granite-
porphyry. Schrader states that the vein is a zone of quartz and
silicified rock from 60 to 80 feet wide, with iron- and man-
ganese-stained croppings. Several shallow prospects have shown
that portions of the vein carry gold.

RUTH VEIN[126]

The Ruth vein is about 1/5 mile south of the Moss mine. Two
claims, the Ruth and Rattan, on this vein have been patented
since 1894. Prior to 1907, it yielded some shipping ore and several
hundred tons of material that was milled at Hardyville. Developments at the Rattan mine include a 200-foot shaft, about 200 feet
of drifts, and considerable stoping on the 100-foot level.

The Ruth vein strikes westward and dips about 80° N. The
ore body is said to be generally less than 2 feet wide and to
consist of calcite, quartz, breccia, and green fluorite. On the 200
level of the Rattan mine, the vein is reported to be one foot
wide, with a one-inch streak of rich ore.

MOSSBACK MINE[127]

The Mossback mine is about 2¼ miles east of the Moss mine,
at an altitude of about 2,400 feet.

This deposit was located in 1863. Prior to 1907, it was developed by a 330-foot shaft in which water rose to within 170 feet

[124] Largely abstracted from Schrader, work cited, pp. 176-78.
[125] Largely abstracted from Schrader, work cited, pp. 179-80.
[126] Largely abstracted from Schrader, work cited, pp. 172-73.
[127] Largely abstracted from Schrader, work cited, pp. 168-70, 1909.

of the surface. During 1918 and 1919, the shaft was sunk to a
depth of 700 feet. Since 1927, considerable development on the
400 level has been carried out by M. Luzon. Early in 1934,
about twelve men were employed, and a small production of concentrates was being made with a pilot mill.

The Mossback vein strikes northwestward and dips 80° SW.,
with a hanging wall of Moss quartz monzonite-porphyry and a
foot wall of Oatman andesite. Its filling, which consists of calcite, quartz, and brecciated andesite, is about 23 feet wide at the
surface. The gold occurs mainly within pale-greenish streaks
and splotches in the quartz.

GOLD ORE MINE

The Gold Ore property, about ¾ mile northeast of the Gold
Road mine, produced bullion at various times from 1918 to 1926,
but no information on the value of the output is available.

Lausen describes this deposit as follows: "The vein occurs
in a fault fissure which strikes N. 68° W. and dips 82° to the southwest. The country rock is all latite, and, along parts of the vein
is intensely altered. At the surface, the altered rock is kaolinized and stained yellow to brown by iron oxides. Although the
altered rock is the same in appearance as the altered andesites, the
presence of abundant flakes of bleached biotite suggests that the
rock was originally latite.

"The shaft is down to a depth of 800 feet and various levels
were run at 100-foot intervals. Most of the ore mined occurred
between the 300-foot level and the 600-foot level. Only a small
amount was found above the 300-foot level, and little mining was
done below the 600-foot level. According to Victor Light, the ore
shoot was approximately 100 feet long and had a width of 4 to 5
feet.

"Several stages of quartz deposition occurred in the vein. Specimens of glassy quartz, coarse grained and stained with copper,
were found. There appeared to have been an abundance of the
well-banded, chalcedonic quartz of the third generation, and such
material was traversed by yellow stringers of quartz with adularia. Some of this yellow quartz contained visible clusters of
free gold and undoubtedly was high grade. Pyrite was common
in the walls of the vein, but, near the surface, it was oxidized "

FUTURE POSSIBILITIES OF OATMAN DISTRICT

Ransome[128] says: "That additional ore bodies remain to be
discovered in the Oatman district at moderate depths is probable.
The district, as has been shown, presents some rather unusual difficulties to prospecting of the ordinary kind, as some of the largest ore bodies thus far discovered gave little or no indication of

---

[128] Ransome, F. L., work cited, p. 56.

their presence at or near the surface."

Lausen[129] gives the following suggestions: "Outcrops which consist only of stringers of chalcedony and calcite may be worth prospecting in the eastern part of the Oatman district where ore shoots which have not been exposed by erosion may exist.

"Excepting in that part of the Oatman district just mentioned, it is useless to prospect veins unless they contain quartz of the fourth or fifth stage of deposition, as described in this report, or unless they show evidence of having been subjected to leaching and supergene enrichment.

"Mineralization has been connected with faulting, and veins which contain crushed zones through which run small stringers of fourth- or fifth-stage quartz and associated minerals offer particularly attractive possibilities.

"Outcrops which show black manganese stains, no calcite, more or less iron stain in the vein or wall rocks, and occasional low gold values should be prospected in the hope of striking a supergene enrichment of gold.

"There is apparently no reason why ore shoots should not be found in any of the igneous rocks that occur in these districts. The wall rocks have suffered little or no replacement, and their effect upon ore deposits has probably been physical rather than chemical. They promoted ore deposition when brittle enough to shatter and provide openings for the passage of solutions and the deposition of minerals.

"Speaking generally, the eastern part of the Oatman district appears to offer greater promise than the western portion because it is believed that ore shoots not yet exposed by erosion may exist there. It may be difficult or impossible, however, to locate veins in the upper latite flows, because they are either very inconspicuous or else differ materially in mineral content from veins which outcrop to the west.

"It should be remembered that ore shoots in the western part of this district have probably suffered more erosion than to the east, and, once found, may be expected to bottom within a few hundred feet.

"Although many veins in the Oatman district have been explored by shafts or drifts at depths of from 100 to 600 feet, such exploration has rarely been continuous for the full known lengths of the veins. The portions still unexplored have some potential value, especially if the fourth- or fifth-stage quartz was found in the parts already mined.

"In spite of the fact that good primary ore has been found at a depth of 1,100 feet below the surface in the Black Eagle mine of the Tom Reed Company, it is not believed that the chance of finding other ore shoots beneath ore already mined and apparently bottomed is good enough to justify deeper exploration."

---

[129] Lausen, Carl, work cited, pp. 124-25.

UNION PASS OR KATHERINE DISTRICT

SITUATION AND ACCESSIBILITY

The Union Pass or Katherine district, as here considered, is some 12 miles north-northwest of Oatman and extends from Union Pass, in the Black Mountains, westward to the Colorado River. (See Figure 6.) Its principal route of access is an improved highway, 35 miles long, that connects the Katherine mill with Kingman. Other roads lead from this highway to the various mines.

HISTORY AND PRODUCTION

The Sheeptrail-Boulevard and a few other properties in this district were discovered during the sixties, but some, as the Katherine, were not found until 1900 or later. Except for the Katherine, which was closed in 1930, most of these mines have long been idle. During 1933-1934, the Roadside, Arabian, and Tyro properties resumed production. The ore from these mines is treated in the Katherine mill, now operated by the Gold Standard Mines Corporation.

The total production of the Union Pass district probably amounts to $3,000,000, of which about 85 per cent was in gold and the balance in silver.

TOPOGRAPHY AND GEOLOGY

For some 2 miles west of Union Pass, the Black Mountains are made up of the Oatman type of Tertiary volcanic rocks, carved into steep, sharply dissected slopes. Farther west, the prevailing rocks consist of a basement of sheared, coarse-grained granite and gneiss of probable pre-Cambrian age, locally overlain by foothill remnants of the volcanic series and intruded by rhyolitic dikes (see Figure 6). Some 4½ miles east of the Colorado River, the foothills give way to a hilly pediment that is largely mantled with gravel and sand.

VEINS

As indicated by Figure 6, the principal veins of the Union Pass district occur within an eastward-trending belt generally less than 2 miles wide. In form, mineralogy, stages of deposition, wall-rock alteration, and origin, they are generally similar to the veins of the Oatman district, described on pages 83-89.

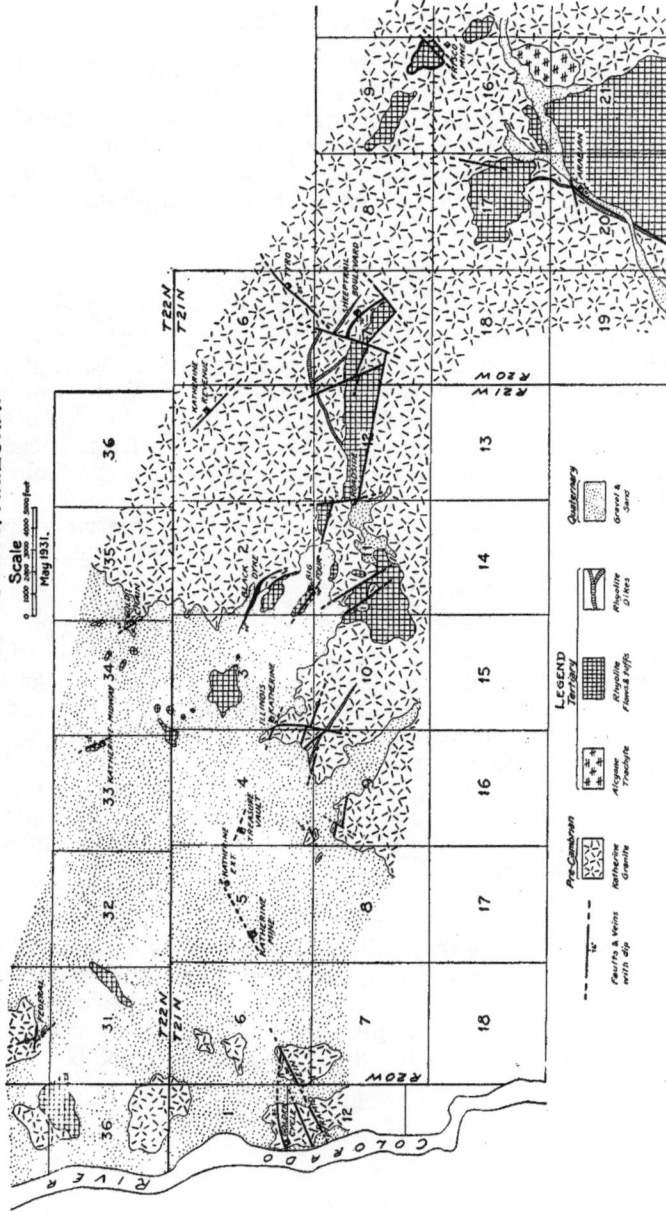

Figure 6.—Geologic map of the Katherine district, by Carl Lausen, 1931.

MINES OF THE UNION PASS DISTRICT

KATHERINE MINE[130]

The Katherine mine is 2 miles east of the Colorado River, in Sec. 5, T. 21 N., R. 20 W.

This deposit was discovered in 1900. The New Comstock Mining Company developed the mine to the 300-foot level and, in 1903, leased the property to the North American Exploration Company which mined out the richer parts of the vein above the 300 level and closed in 1904. The Katherine Gold Mining Company sank a new shaft, built a 150-ton cyanide mill, and operated from 1925 until late 1929. Lessees made a small production from ore and old tailings during 1930 and 1931. The mine has been closed since 1930.

APPROXIMATE PRODUCTION OF KATHERINE MINE, 1925-1931.

(Data compiled by J. B. Tenney)

1925	$ 200,000
1926	400,000
1927	300,000
1928	150,000
1929	25,000
1930	9,400
1931	2,600
Total	$1,087,000

Of this total value, about 85 per cent was in gold and 15 per cent in silver.

In 1933, the Katherine mill and the rights to the water of the Katherine mine were purchased by E. F. Niemann and associates who later formed the Gold Standard Mines Corporation. Early in 1934, this Company was employing about forty men. The mill, which has a capacity of 300 tons, was treating about 60 tons per day. Most of this ore was from the Roadside and Arabian mines (see pages 105-106), and part was customs ore.

The Katherine mine is on a small knob of granite, about 150 feet across, that rises slightly above the general level of the surrounding gravel-floored plain. The collar of the shaft is 990 feet above sea level, or 450 feet above the Colorado River, and the water table is at approximately the 350-foot level.

The vein strikes N. 62° E. and dips about vertically. It is a stringer lode that has a width of more than 60 feet at the surface but narrows underground. This lode has been opened by a 900-foot shaft and for a length of 1,700 feet. Lausen described the vein as follows: "Vein filling at the Katherine mine usually consisted of a series of closely spaced stringers in the granite. At

---

[130] Largely abstracted from Lausen, Carl, Geology and ore deposits of the Oatman and Katherine districts, Arizona: Univ. of Ariz., Ariz. Bureau of Mines Bull. 131, 1931.

some places, however, the vein filling was solid quartz and calcite up to 10 or more feet wide. Much of the vein filling consisted of quartz, but here and there calcite was abundant. Various stages of quartz deposition are represented at this mine, but the first and second are the most abundant . . . An intergrowth of quartz and adularia also occurred here and formed some of the important ore shoots. It was very similar in appearance to that occurring at Oatman. Sometimes the adularia was rather coarse grained, and the associated quartz was of a deep greenish yellow color. Platy quartz, white in color, with a well-developed laminated structure, also occurred in the Katherine vein. Some of the smaller stringers frequently showed a fine banding and were usually frozen to the somewhat silicified granite walls.

"At the west end of the 200-foot level, some rich silver ore was mined, and, on the 900-foot level, a small stringer, from 2 to 6 inches wide, assayed 65 ounces in silver and 10 ounces in gold. Copper stain was abundant in this rich stringer, associated with chalcocite."

Near the vein, the granite is kaolinized, iron-stained, and locally silicified.

The vein has been cut by numerous faults that have been interpreted as low-angle thrusts from the southwest.[181] The maximum offsetting of an ore body in the plane of the vein is 60 feet. Later movements on the vein have cut the thrust planes.

Many details on mining, milling, and costs are given in the article by Dimmick and Ireland.

ROADSIDE MINE

The Roadside mine, as shown on Figure 6, is 4 miles east of the Gold Standard mill.

During 1915-1916, the present shaft was sunk to the 100-foot level, and some drifting was done. Further exploration was carried on in 1921. In 1932, E. Ross Householder leased the mine. Later in the same year, E. F. Niemann and associates, who subsequently formed the Gold Standard Mines Corporation, obtained control of the property. Up to January, 1934, this company had done about 1,000 feet of development work on the 100 level and mined ore that, treated in the Gold Standard mill (see page 103), yielded about 890 ounces of gold and 1,734 ounces of silver.

Here, a fan-shaped block of rhyolite, some 32 feet wide at its narrowest or southern margin, is in fault contact with granitic gneiss on the east and west. The Roadside lode strikes northward, dips 33° to 38° W., and occurs within a fault zone in the rhyolite. The ore shoot, as exposed underground is from 20 to 35 feet wide and 75 feet long on the strike. It consists of irregular stringers and bunches of quartz and calcite in shattered, silici-

---

[181] Dimmick, R. L., and Ireland, E., Mining and milling at the Katherine mine: Eng. and Min. Jour., vol. 123, pp. 716-20, 1927.

fied rhyolite. The gold occurs mainly in greenish-yellow quartz and also in streaks of chocolate-brown iron oxide. According to E. F. Niemann,[132] the ore shoot, where explored between the 100-foot level and the surface, averages from 0.25 to 0.30 ounces of gold to the ton.

ARABIAN MINE

The Arabian mine is in Sec. 20, T. 21 N., R. 20 W., about 3½ miles southeast of the Roadside property.

Intermittent work, which resulted in a small production, has been carried on at this property since before 1917. In late 1933 and in 1934, E. F. Niemann and associates worked the mine and milled several thousand tons of the ore in their Gold Standard plant. The 1933 production amounted to about 593 ounces of gold and 1,156 ounces of silver.

Lausen says: "At this mine, a rhyolite-porphyry dike intrudes granite, and, along the hanging wall of this dike, the rhyolite tuffs have been faulted against the dike. The vein occurs in the dike, close to the fault, and it strikes northeastward while the dip is 82° to the southeast.

"A mineralized zone, 30 feet wide and consisting of a number of quartz stringers, occurs in the rhyolite dike and, to a certain extent, in the granite footwall. The individual veinlets of this zone vary in width from a fraction of an inch up to 12 inches or more. The veinlets are chiefly quartz, but, in some places, consist of coarse-grained, gray calcite. A comb structure is common in the smaller stringers where the quartz crystals are large. The central portion may be vuggy, and the vugs often contain manganese dioxide; occasionally, however, the central part of the veinlet is filled with calcite. Near the hanging wall portion of the lode, a small stringer of fluorite was found. Near the portal of the tunnel is some waxy, yellow quartz, a part of which had replaced calcite. No adularia was found in this quartz, but the best values occur in this portion of the lode."

This mineralized zone forms an outcrop 100 feet high by more than 100 wide and more than 500 feet long. Preliminary sampling by Mr. Niemann indicates that much of it may carry more than 0.10 ounces of gold to the ton. Early in 1934, it was being mined by open cut methods and sent to the Gold Standard mill.

Near the northern end of the property, a vein about six feet in maximum width strikes northward, dips 33° E., and is separated from the big lode by faulting. This vein contains an ore shoot that, early in 1934, was being mined for a length of a few tens of feet between the 80-foot level and the surface. According to Mr. Niemann, this shoot averages about 0.50 ounces of gold and from 3 to 10 ounces of silver to the ton.

---

[132] Oral communication.

TYRO MINE

The Tyro mine is in Sec. 6, T. 21 N., R. 20 W., and 6.1 miles by road east of the Gold Standard mill.

During 1915 and 1916, the Tyro shaft was sunk to a depth of 500 feet, and some drifting was done on the 200-foot level. Some ore was produced from small pockets near the surface. During 1933-1934, W. E. Whalley and C. F. Weeks, lessees, built a road from the mine to the Katherine highway and began production from surface cuts on the vein.

Here, coarse-grained gneissic granite, cut by numerous narrow dikes of rhyolite-porphyry, forms rugged topography. The vein strikes northeastward, dips 85° SE., and forms a stringer lode with a prominent outcrop some 1,800 feet long by 20 to 35 feet wide. The stringers, according to Lausen, consist mainly of granular white quartz with platy calcite and, in places, glassy, yellowish quartz of probably the second stage of deposition. He states that the vein was not found in the deeper workings of the mine.

SHEEPTRAIL-BOULEVARD MINE

The Sheeptrail-Boulevard mine is in Sec. 7, T. 21 N., R. 20 W., about 7 miles east of the Colorado River.

This deposit, according to Schrader,[133] was discovered in 1865. It was acquired by the New Comstock Mining Company which treated about 2,000 tons of the ore in a 20-stamp mill at the river. The Arizona-Pyramid Gold Mining Company acquired control of the property in 1904 and milled considerable ore. The total production is estimated at 15,000 tons.

Here, granite and rhyolite form a group of low hills. Lausen[134] describes the deposit as follows: "The vein occurs near the contact of granite with a dike of rhyolite-porphyry which forms the hanging wall at the west end of the mine. Small stringers occur both in the granite and in the dike, but most of the ore mined appears to have come from the rhyolite. The vein strikes northwest, dips south, and takes a curved course, trending more nearly east-west towards the northwestern end. At the west end, it is cut by a northeast fault. A number of minor faults which trend northeast cut the vein, but, in each case, the offset is small.

Mineralization consists of a number of small stringers of quartz over a width of from 3 to 7 feet. This quartz is not everywhere ore, and only certain portions of the vein stained with iron oxides were mined. Much of the quartz is fine grained, and some of it shows a platy structure.

"The mine was developed to a depth of 450 feet by an inclined shaft, and considerable drifting was done from this shaft. At the surface, numerous tunnels have been driven into the vein.

[133] U. S. Geol. Survey Bull. 397, pp. 204-205.
[134] Work cited, p. 120.

The water table occurs 40 feet below the collar of the shaft, and no ore was mined below the 350-foot level."

FRISCO MINE[135]

The Frisco Mine is in the eastern part of the area shown on Figure 6, about 9 miles east of the Colorado River.

This deposit, which was located about 1900, made an estimated production of 44,000 tons of ore prior to 1916. In 1932, according to E. Ross Householder,[136] it yielded a considerable tonnage of ore that was treated in the Katherine mill.

Here, granite, capped by rhyolite, forms a low hill. The principal vein strikes N. 55° E., dips 12° SE., and occurs as a stringer lode in the rhyolite at the granite contact. Its maximum width is 18 feet. Several faults, generally of small displacement, cut the vein. The quartz of these stringers is banded, vuggy, and chalcedonic in texture and creamy white to light brown in color. Most of the ore mined came from the heavily iron-stained lower portion of the vein. Some of the underlying iron-stained granite contained ore.

In the granite flat southeast of the hill is a lode that strikes northeastward, dips 65° NW., and is 59 feet in maximum width. As explored by a 300-foot shaft, with drifts on the 200 level, it was not of commercial grade.

BLACK DYKE GROUP[137]

"The Black Dyke group of claims is 3 miles to the east of the Katherine Mine. This large vein is composed principally of calcite, cut by a great number of small stringers of quartz. The vein takes a curved course, trending northwest, and with a length of one-half mile. About midway between the ends it swells to a width of 150 feet . . . The dark color is confined to a thin film at the surface, usually termed 'desert varnish.' The vein appears to occur in a shattered rhyolite which may have been replaced to a certain extent by both the calcite and quartz.

"This vein is said to have been thoroughly sampled and found to average $3 per ton, but the highest assay obtained by the writer was $2.40. A small inclined shaft has been put down at the west end, and numerous small tunnels have been run beneath the outcrop."

PYRAMID MINE[138]

"The Pyramid Mine is probably the oldest location in the district. It is situated near the Colorado River in some low hills of granite. The vein, which consists of a large number of small stringers in reddish granite, strikes N. 65° E. and the dip is ver-

---

[135] Largely abstracted from Lausen, work cited, pp. 121-22.
[136] Oral communication.
[137] Quoted from Lausen, work cited, p. 119.
[138] Quoted from Lausen, work cited, p. 118.

tical. The vein has been prospected by a number of small pits and a shaft at the west end, which is said to have been sunk to a depth of 70 feet. Some rich ore is said to have been stoped from this shaft, but the workings are now caved. No record of production is available."

GOLDEN CYCLE MINE[139]

"About 2,000 feet to the northwest of the Pyramid Mine is the Golden Cycle. The vein there also occurs in the coarse-grained granite. The vein takes a more easterly course than the Pyramid vein, and appears to join this vein somewhat less than a mile to the east. The vein consists of a series of quartz stringers occupying a shear zone in the granite. The general dip of this zone is 82° to the north, and the width is from 10 to 18 feet.

"At the west end of the property, a shaft was sunk to a depth of 115 feet, and some lateral work has been done at this level. Near the surface, samples assay from $1 to $3 per ton, but, underground some samples have assayed $14 per ton."

OTHER PROPERTIES

The *San Diego, O. K. Expansion, Union Pass, Monarch, New Chance,* and *Sunlight* properties are described by Schrader.[140] More or less prospecting has been done on the *Gold Chain, Burke. Mandalay, Bonanza, Banner, Tin Cup* and *Quail* properties.

FUTURE POSSIBILITIES

That additional bodies of gold ore will be discovered in the Union Pass district seems possible, but, as in the Oatman district, the surface indications are inconclusive. The known ore shoots are of the epithermal type and probably do not extend to depths of more than a few hundred feet. At 1934 prices, however, some of the large low-grade deposits of the district appear to offer possibilities for considerable production.

MUSIC MOUNTAIN DISTRICT[141]

The Music Mountain district is in the foothills of the Grand Wash Cliffs, some 25 miles north of Hackberry. Gold deposits were discovered here about 1880. Prior to 1904, the district yielded some $20,000 worth of bullion. Since that time, intermittent operations have resulted in a small production.

In this vicinity, the principal rocks are granite, schist, and gneiss, intruded by dikes of diabase and granite porphyry.

The gold deposits consist of steeply northeastward-dipping quartz veins. They contain considerable iron oxide to depths of about 200 feet, below which pyrite and galena are locally abun-

---

[139] Quoted from Lausen, work cited, p. 119.
[140] U. S. Geol. Survey Bull. 397, pp. 208-14.
[141] Abstracted from Schrader, work cited, pp. 142-49.

dant. In general, their ore shoots are only a few inches wide but of high grade.

ELLEN JANE MINE[142]

The Ellen Jane mine is on the Ellen Jane vein, near the head of Camp Wash, at an altitude of about 3,440 feet.

This vein was located in 1880. In 1886, the Gold Mining Company purchased it from W. F. Grounds and erected a small mill in which about 700 tons of ore was treated. Some years later, the Washington-Meridian Company operated the property but closed in 1904. Since that time, a small production has been made by lessees.

The vein strikes northwestward, dips 80° NE., and occurs associated with a diabase dike in granite. Its gangue, which is brecciated quartz with altered country rock, ranges from 2 to 7 feet in width. The ore streak averages only about 5 inches in width and generally occurs near the hanging wall. It is oxidized to the 200-foot level below which some pyrite and galena are present. In places, the ore was of high grade.

In 1907, underground developments at the Ellen Jane mine included a 200-foot shaft and about 1,500 feet of workings. The shaft was making about five barrels of water per day.

OTHER PROPERTIES

The *Mary E* and *Southwick* veins are briefly described by Schrader.[143] Prior to 1907, they had been developed to shallow depths and had yielded a few carloads of rich ore.

The *Rosebud* claim, near the mouth of the basin, was located shortly after the Ellen Jane vein. From about 1926 to 1932, it was worked by the Portland and Mizpah Mining Company. According to the U. S. Mineral Resources, this company produced some bullion with a 100-ton mill in 1929, but the recovery was poor. In 1931, with a new 20-ton amalgamation and cyanide mill, about $4,000 worth of bullion was produced from 500 tons of ore. Since 1932 little work has been done on the property. The underground developments are reported to include a 400-foot shaft and some 2,500 feet of workings.

CERBAT MOUNTAINS

The Cerbat Mountains are in west-central Mohave County, some 20 miles east of the Colorado River. From the Santa Fe Railway, at Kingman, they extend north-northwestward for 30 miles, with a maximum width of 12 miles. They attain an elevation of 7,000 feet above sea level or 4,000 above the adjacent desert plains.

Kingman and Chloride are the principal towns in this area.

---

[142] Abstracted from Schrader, work cited, pp. 144-48.
[143] Work cited, pp. 148-50.

A branch railway, built in 1899 to connect Chloride with the Santa Fe, was recently dismantled.

These mountains consist largely of pre-Cambrian schist, gneiss, and granite, intruded by granite-porphyry and lamprophyric dikes, and overlain in places by Tertiary volcanic rocks.

The principal mineral deposits are in the west-central segment of the range, in the Chloride, Mineral Park, Cerbat, and Stockton localities, which are collectively termed the Wallapai or Hualpai district. The deposits are mesothermal veins of prevailing northwestward strike and steep dip. Their gangue[144] is quartz, in many places shattered and recemented by later calcite. The primary minerals include pyrite, chalcopyrite, arsenopyrite, galena, sphalerite, tennantite, proustite, and pearceite. Locally, gold occurs in the sulphide zone. In the oxidized zone are native silver, horn silver, ruby silver, oxidized lead minerals, and locally, native gold. Rich bodies of silver ore with some gold were found in the oxidized zone, but sphalerite seems to be the principal constituent of the sulphide zone. The water level is generally above depths of 400 feet. Silver and lead predominate in the Chloride, Mineral Park, and Stockton localities, and gold and silver in the vicinity of Cerbat camp.

The Cerbat Mountains have made a large production in silver, lead, zinc. and gold. Only those deposits that have been valuable chiefly for gold are considered in this report.

CHLORIDE VICINITY

PAY ROLL MINE[145]

The Pay Roll mine is about 1½ miles east of Chloride, near the middle of the western slope of the range. This deposit was located in 1887 and, prior to 1907, was opened by shafts 225, 100, and 60 feet deep, with 400 feet of workings and 600 feet of tunnels. The production was reported to include many carloads of shipping ore that ran about $80 per ton, mostly in gold, and was derived principally from the surface workings. Some development work was carried on intermittently from 1916 to 1929, and a small production of ore containing zinc, lead, silver, and gold was reported in 1917. In 1929, the Pay Roll Mines, Inc., treated about 1,400 tons of complex copper-lead-zinc ore in their new 50-ton flotation plant, but ceased operations late in the year.

The mine is on the Pay Roll vein, which strikes N. 30° W., dips steeply northeastward, and occurs in schist. This vein ranges from 6 to 100 feet, with an average of about 10 feet, in thickness and is traceable on the surface for nearly a mile. Its gangue is mainly quartz. The ore minerals are lead carbonate, galena, pyrite, and chalcopyrite, with both gold and silver.

[144] Description abstracted from Schrader, F. C., U. S. Geol. Survey Bull. 397, 1909.
[145] Largely abstracted from Schrader, work cited, pp. 62-63.

RAINBOW MINE[146]

The Rainbow mine is about 2½ miles east of Chloride, at an altitude of about 5,500 feet.

This deposit was discovered in 1883 and worked intermittently until 1890. In 1907, it was being worked by the Rainbow Mining Company. Some production was reported in 1911, 1913, 1916, 1920, and 1933. The production of the mine amounts to about $75,000, most of which was made prior to 1891.

The Rainbow vein strikes north-northeastward dips vertically and occurs in silicified gneissoid granite. Several spur veins join the main vein and form ore shoots at the junctions. The ore streaks are persistent and range from 6 inches to 3 feet in width.

The principal ore shoot was developed by a 260-foot drift and largely worked out to the sulphide zone. This shoot is about 12 inches wide by 150 feet long.

Some 200 feet farther north is the Windsail ore shoot which contains a streak of quartz and cerussite 15 inches wide by 25 feet long. It was developed by a 125-foot shaft with about 1,000 feet of lateral workings.

Some forty-five assays from various parts of the Rainbow workings showed an average of 2 ounces of gold and 20 ounces of silver per ton, and about 12 per cent of lead.

SAMOA MINE[147]

The Samoa mine is 3½ miles east of Chloride, at an elevation of 5,900 feet.

For many years, the Samoa was the most active and constant producer of gold and silver ore in the Chloride area.

Prior to 1903, its yield amounted to approximately $70,000 From 1903-1908, under the ownership of the Chloride Gold Mining Company, its production amounted to approximately $110,-000. In 1907, it was sending about 90 tons of ore per month to the Needles smelter. All of the ore was packed by burros to the railway at Chloride. Since early 1908, only a small, intermittent production, by lessees, has been made. In 1931, the property was acquired by the Samoa Gold Mines Corporation.

Here, the prevailing rocks consist of gray granite, intruded by a large dike of microcline granite and a 100-foot dike of rhyolite-porphyry.

The principal vein strikes N. 10° W., dips about 80° E., and averages 4 feet in thickness. In 1907, at the time of Schrader's field work, it had been explored by a 335-foot shaft and more than 3,000 feet of workings. Most of the ore mined was from above the second level. The ore shoot was 30 inches in maximum width and more than 800 feet long It contains pyrite, galena.

---

[146] Abstracted from Schrader, work cited. pp. 64-67.

[147] Abstracted from Schrader, work cited, pp. 67-69.

and sphalerite in quartz gangue, and a little molybdenite in carbonate veinlets. Some specimens show sphalerite coated with black silver sulphide, and others contain considerable native silver. As shown by smelter sheets, the ore shipped during 1903-1906 averaged 8 per cent of lead, 5 to 8 per cent of zinc, 15 ounces of silver, and 1½ ounces of gold per ton. Schrader states, however, that the mine as a whole is a low-grade property.

TINTIC MINE

The Tintic mine, on the pediment about 1½ miles west of Chloride, is reported to have produced several thousand dollars' worth of gold ore prior to 1907. A little development work was done in 1918, and some production was reported in 1927, 1928, 1931, and 1932. In 1933, Rae Johnston and associates obtained an option on the property.

The vein dips gently northeastward and is from 2 to 10 feet[148] thick. Most of the ore mined came from shallow depths and within a horizontal distance of about 200 feet.

MINERAL PARK VICINITY

TYLER MINE[149]

The Tyler mine is 2¾ miles southeast of Mineral Park or 6¾ miles southeast of Chloride, at an altitude of about 5,300 feet. It was discovered in 1901 and shipped ore from 1905-1907.

The vein strikes N. 37° W., dips 75° SW., and occurs in gneissic granite. It is about 40 feet wide and consists mainly of altered, brecciated aplite, locally cemented by quartz. The ore, which occurs mainly near the hanging wall, consists principally of galena and cerussite. The last carload shipped at the time of Schrader's visit averaged 17.5 per cent of lead, 8 ounces of silver and 3.16 ounces of gold per ton.

CERBAT VICINITY

GOLDEN GEM MINE[150]

The Golden Gem mine is at Cerbat settlement, near the western foot of the range, about 8 miles southeast of Chloride.

This deposit was discovered in 1871 and was successively worked by T. L. Ayers, the Golden Gem Mining Company, and the Golden Star Mining and Milling Company. Little underground work has been done since 1904. In 1907, at the time of Schrader's visit, the Golden Star Mining and Milling Company was producing, with a 40-ton mill, about $350 worth of concentrates daily.

---

[148] Description abstracted from Schrader, work cited, p. 79.
[149] Abstracted from Schrader, work cited, pp. 82-83.
[150] Abstracted from Schrader, work cited, pp. 92-94.

Here, the prevailing rocks are schist, gneiss, and granite, intruded by porphyritic dikes of acid to basic composition.

The vein strikes N. 40° W., dips 78° NE., and is traceable for about 1/5 mile southeastward. It ranges in width from 6 to 14 feet. The gangue is imperfectly banded quartz with included fragments of country rock. The ore minerals are galena, sphalerite, pyrite, and stibnite. The ore contains chiefly gold with locally abundant silver.

Underground workings include a 435-foot shaft with about 1,200 feet of workings. As shown by these workings, the ore is from 2 to 6½ feet wide. The only stopes are on the 130-foot level. According to Schrader's sketch, they extend, with a height of 62 to 81 feet, for 266 feet northwest of the shaft, to the vicinity of a fault that cuts off the ore. On the 300-foot level, the ore was of low grade.

IDAHO MINE[151]

The Idaho mine, a short distance west of the Golden Gem property, was worked for gold prior to 1871. In 1907, its workings included a 110-foot shaft and 275 feet of drifts. The vein strikes N. 20° E., dips 80° E., and averages about 4 feet in width. Its ore shoot, which is about 2 feet wide, contains galena, pyrite, a little chalcopyrite, and some silver sulphide.

VANDERBILT MINE[152]

The Vanderblit mine, about ½ mile northwest of the Golden Gem, was located in the sixties and was producing in 1907, at the time of Schrader's visit. Most of the ore mined came from the upper levels.

The vein strikes northwestward, dips 80° NE., and occurs in altered chloritic schist. Developments on the vein include a 300-foot shaft and 800 feet of drifts. The vein, like that of the Golden Gem mine, contains quartz, pyrite, galena, sphalerite, and stibnite. The ore, which is valuable mainly for gold and silver, is prevailingly of low grade. Lead-bearing ore occurs only above the 200-foot level.

FLORES MINE[153]

The Flores mine, in Flores Wash, about ½ mile northwest of the Golden Gem property, was discovered in 1871, and worked to some extent during the late seventies. It was operated from 1888 until 1893 by the Flores Mining Company and then sold for taxes. In 1915, La Anozira Mining Company made a small production from the property. In 1907, workings on the property included a 300-foot shaft with but little stoping.

The vein which occurs in granitic schist strikes northwestward,

---

[151] Abstracted from Schrader, work cited, p. 95.
[152] Abstracted from Schrader, work cited, pp. 94-95.
[153] Largely abstracted from Schrader, work cited, p. 96.

dips almost vertically, and is about 4 feet thick. Its croppings are chiefly iron-stained quartz and breccia. The ore contains free gold, silver sulphide, sphalerite, and galena and occurs mainly near the hanging wall of the vein. Some of the ore mined during the early days was of high grade, but the material milled by the Flores Company ran as low as $6 per ton.

ESMERALDA MINE[154]

The Esmeralda mine is in the foothills about one mile northwest of the Golden Gem property. Its vein, which occurs in schist, strikes N. 35° W., dips 75° SW., and is 4 to 5 feet wide. The gangue is quartz and altered breccia. Oxidation extends to a depth of 90 feet, below which the ore minerals consist mainly of pyrite and chalcopyrite with gold and silver. One carload of the concentrates is reported to have averaged 13.42 ounces of gold and 40 ounces of silver per ton. Underground developments include a 200-foot shaft, about 200 feet of drifts, and some stoping near the surface.

CERBAT MINE[155]

The Cerbat mine is about one mile northeast of the Golden Gem property, at an altitude of 4,600 feet.

This deposit was worked to a considerable extent during the early days, and was equipped with a mill about 1880. The property has long been inactive.

The vein strikes north-northwestward, with bold croppings 4 to 10 feet wide. Its gangue consists of quartz with some brecciated country rock. The ore contains chiefly gold with some silver and copper. Underground workings include a 180-foot shaft with some drifts and stopes.

ORO PLATA MINE[156]

The Oro Plata mine is east of the Cerbat mine, at an altitude of about 4,300 feet.

This deposit was worked by Mexicans during the early seventies. Later, it was operated by J. P. Lane, by H. Wilson after 1882, and by J. W. Garret after 1895. During 1897-1898, its production amounted to $150,000. As shown by sheets of the Arizona Sampler Company, which bought the ore at Kingman, the mine produced, from 1896 to 1901, 2,527 tons of ore that averaged $80 per ton. Approximately 75 per cent of this value was in gold and the balance in silver and lead. About 1906, the Oro Plata Mining Company purchased the property. The total production of the mine is reported to have been worth $500,000.

Here, gneiss has been extensively intruded by granite porphyry. The vein strikes N. 55° W., dips 80° NE., and is about 4 feet wide.

---

[154] Abstracted from Schrader, work cited, p. 97.
[155] Abstracted from Schrader, work cited, pp. 97-98.
[156] Abstracted from Schrader, work cited, pp. 100-101.

Its ore, which occurs mainly in banded quartz, contains principally gold and silver, with some chalcopyrite, sphalerite, pyrite,
and galena. The mine is developed to a depth of 280 feet by
shafts, drifts, adits, and stopes that aggregate about 7,000 feet.

COTTONWOOD DISTRICT

WALKOVER MINE

The Walkover mine is in the northern portion of the Cottonwood Cliffs plateau, about 9 miles by road east of Hackberry.
This mine produced intermittently from about 1911 to 1918.
Sheriff sales of the Walkover Mining Company's property were
held in 1921 and 1923. Lessees made a small production in 1925.
In 1927, the Walkover Mines, Inc., was formed and later changed
to the Calizona Mining Company. In 1928, El Paloma Mining
Company was formed. During 1928 and 1929, a little bullion and
two cars of shipping ore were produced. A 30-ton concentration-
cyanidation mill, installed early in 1930, proved unsuccessful. In
1933, the property was leased to F. Nielson and associates who,
up to February, 1934, had shipped four carloads of ore that contained from 1.00 to 1.29 ounces of gold per ton.[157]
The Cottonwood Cliffs represent the southeastward continuation of the Grand Wash Cliffs and mark the western limit of a
plateau that here attains an elevation of about 5,000 feet above
sea level or 1,300 feet above Hackberry. This portion of the
plateau consists mainly of granite and schist overlain by essentially horizontal Tertiary volcanic rocks.
The Walkover vein strikes S. 25° E., dips 75° W., and occurs in
mediumly foliated dark-colored schist. It has been explored to a
depth of 365 feet and for a maximum length of 200 feet south of
the main shaft, to where it is cut off by a fault that strikes S.
60° W. Thirty feet south of this shaft, the vein is displaced 75
feet westward by a fault that strikes S. 70° W. and dips 80° S.
The gangue of the vein is brecciated, glassy, gray quartz and
altered fragments of schist. Above the 100-foot or water level,
it was accompanied by considerable iron oxide and some copper
stain. Below that level, it contains abundant banded sulphides,
mainly pyrite, arsenopyrite, and chalcopyrite.
The vein has been largely stoped from the surface to the 100-
foot level. As indicated by these openings, the width of the ore
shoot ranged from 8 inches to 5 feet and averaged about 2 feet.

CHEMEHUEVIS DISTRICT

The Chemehuevis district is in the Mohave Mountains of southwestern Mohave County. These mountains constitute a rugged
range, about 34 miles long by a maximum of 12 miles wide, that
trends southeastward from the Colorado River at a point a few

[157] Oral communication from F. Nielson.

miles south of Topock. The highest peak has an altitude of
about 4,300 feet, but most of the mountains are considerably
lower. The region as a whole is very arid. Sandy roads from
Topock and Yucca, on the Santa Fe railway, lead around the base
of the range.

In their main or northeastern portion, these mountains are
made up of banded schist, intruded by dikes of diorite and granite-porphyry. The southwestern flank of the range consists of
generally westward-dipping volcanic rocks.

The typical gold deposits of the Mohave Mountains are veins
of coarse-textured brecciated white quartz in schist. Their gold
occurs mainly in association with pyrite and galena that show but
little oxidation near the surface. The veins are locally of high
grade, but tend to be narrow and pockety. During the present
century, they have produced generally less than a thousand dollars' worth of gold per year.

The *Best Bet* or *Kempf* property is in the southwestern portion
of the range, about 55 miles by road from Topock. It was located during the early days and held for several years by A.
Kempf. Early in 1933, Mrs. Isabella Greenway obtained control
of the ground, built many miles of new road, and installed a 50-
ton flotation mill at the Colorado River, 12 miles west of the
mine. According to local reports, the mill treated about 540 tons
of $15 gold ore from the Kempf mine, some 300 tons of silver-
lead ore from the Mohawk mine, and a little gold ore from the
Santa Claus and other properties. Operations were discontinued
in the summer of 1933, and the property reverted to Mr. Kempf.
At the Best Bet mine are a few veins that strike northeastward,
transverse to the schistosity, and dip steeply southeastward. As
seen in several shallow shafts and short tunnels, their widths
range from less than an inch up to about 1½ feet and average
only a few inches. According to local reports, the mined portion
of the principal vein was about 30 inches in maximum width.

The *Gold Wing* property, a few miles north of the Best Bet, has
produced a small tonnage of ore that was treated in a 3-stamp
mill near the Colorado River. The *Susan* and *Santa Claus* prospects, also in this vicinity, have recently yielded a small production.

The *Dutch Flat* mine, on the opposite side of the range, for
many years has produced a small tonnage of ore that was treated
in a 10-ton mill on the property.

CHAPTER IV—COCHISE COUNTY

Cochise County, as shown by Figure 12 (page 186), comprises
an area about 80 miles long by 75 miles wide. It consists of wide
plains surmounted by large mountain ranges of complexly faulted pre-Cambrian schist and granite, Paleozoic and Cretaceous

sedimentary beds, Cretaceous and Tertiary intrusives, and Tertiary volcanic rocks.

This county, which ranks third among the gold-producing counties of Arizona, to the end of 1931, produced approximately $30,-230,000 worth of gold. Of this amount, about $25,475,000 worth was a by-product from copper ores, mainly from the Bisbee and Courtland districts, and $273,500 worth was a by-product of lead mining.[158]

DOS CABEZAS MOUNTAINS

The Dos Cabezas Mountains, of northeastern Cochise County, constitute a northwestward-trending range, about 20 miles long by 3 to 10 miles wide, with a maximum altitude of 8,300 feet above sea level or more than 4,000 feet above the adjacent plains. Its principal settlement, Dos Cabezas, is at the southwestern foot of the range, 15 miles by road and branch railroad from Willcox, a station on the Southern Pacific Railway.

The range is made up of pre-Cambrian schist and granite, Paleozoic to Mesozoic sedimentary beds, Tertiary volcanic rocks, and Mesozoic or Tertiary intrusives of acid to basic composition. These formations have been affected by complex faulting of both normal and reverse character.

Gold deposits occur in the vicinity of Dos Cabezas; in upper Gold Gulch, northwest of Dos Cabezas; in the Teviston district, at the northern end of the range; and in the vicinity of Apache pass, at the southeastern end of the range. These deposits were discovered prior to the Civil War and have been worked intermittently since the seventies. Up to 1933, they yielded approximately $182,000 worth of gold.[159] Most of this production was made by the Dos Cabezas district, which has also yielded notable amounts of copper, lead, and silver. During the past few years, the Dives, Le Roy, Gold Ridge, and Gold Prince gold mines have been actively worked. According to the U. S. Mineral Resources, the gold production of the district amounted to $3,841 in 1930, $11,132 in 1931, and $33,901 in 1932.

In the Dos Cabezas gold district, the formations outcrop in westward-trending belts with pre-Cambrian granite on the south, succeeded northward by steeply dipping, metamorphosed Cretaceous shales and sandstones and Carboniferous limestones. These rocks are intruded by dikes of rhyolite-porphyry and diabase. The thrust-fault zone that separates the Cretaceous strata from the granite contains a vein of coarse-grained white quartz, called the "Big Ledge," that attains a maximum width of 100 feet but locally branches and pinches out. In places, this vein carries a little gold, but most of it is of very low grade.

The gold-bearing veins consist of coarse-textured, white to grayish-white quartz with scattered small bunches and dissem-

---

[158] Statistics compiled by J. B. Tenney.
[159] Figures compiled by J. B. Tenney.

inations of galena, pyrite, sphalerite, and chalcopyrite. The gold occurs within the sulphides, mainly the galena, and is free milling only near the surface. Some of the ore is very rich, but most of it contains less than 0.5 ounce of gold per ton.

The veins occur both in the Cretaceous rocks and associated with diabase dikes in the granite. They are of mesothermal type, but tend to be of lenticular form, particularly in the shales. Their walls show silicification. The shales adjacent to the veins contain pyrite metacrysts and abundant graphite.

DIVES MINE

The Dives deposit, about 2½ miles by road north of Dos Cabezas and south of the Central Copper mine, was located in 1877 as the Bear Cave claim. During the eighties, some of its ore was treated in a stamp mill at Dos Cabezas. During 1911-1914, the mine yielded more than $20,000 worth of gold. In 1919, the property was acquired by the Dives Mining Company which erected a 10-stamp amalgamation-concentration mill on the property and operated actively for a few years. The total production of the mine from 1877 to 1920 is estimated at about $40,000 worth of gold from 5,000 tons of ore.[160] About 1922, the Twin Peaks Mining Company took over the property but made little or no production. During 1931, lessees are reported to have mined more than $5,000 worth of ore from the old workings. In 1932, the present operators, *Consolidated Gold Mines Company,* began work on the property. This Company drove the lower tunnel workings and, in May, 1934, completed a 50-ton flotation mill About thirty-one men are employed.

The gold-bearing veins occur within steeply northward-dipping, metamorphosed, black Cretaceous shales that occupy a belt approximately 1,100 feet wide. The "Big Ledge," a vein cf coarse-grained white quartz, 100 feet wide, separates the shales from the granite and, according to local reports, in places contains up to 0.10 ounce of gold per ton.

The older, upper workings include a 100-foot inclined shaft and 1,000 of adit that extends N. 68°-87° W. on a vein that dips almost vertically.[161] This vein, in the first 575 feet of the adit, attains a maximum width of 5 feet but locally pinches out. In the last 425 feet of the adit, its width ranges from one to 10 feet and averages about 4 feet. Some stoping was done at a point about 700 feet in from the portal, on a steeply westward-pitching ore shoot that was from less than one foot to 6 feet wide by 10 to 50 feet long. This adit has a maximum of about 280 feet of backs.

The lower, recent workings, which are 212 feet below the upper tunnel, include an adit that, in June, 1934, extended northward

---

[160] Historical data from unpublished notes of M. A. Allen, 1920.
[161] Unpublished notes of Carl Lausen, 1923.

for 1,800 feet, through the "Big Ledge" and nearly through the shale belt. In the shales, it cut four separate, westward-trending veins that have been opened by drifts from 35 to 360 feet long. As exposed, these veins range from less than one foot to 12 feet wide.

The Dives veins consist of coarse-textured white quartz with scattered bunches and disseminations of galena, pyrite, sphalerite, and chalcopyrite. Their gold occurs in the sulphides, mainly the galena, and is free milling only near the surface. According to A. B. Wadleigh,[162] Superintendent of the Consolidated Gold Mines Company, the ore carries from one to 2 ounces of silver per ounce of gold. Graphite is abundant in the wall rock.

GOLD RIDGE OR CASEY MINE

The Gold Ridge or Casey property, controlled by Mrs. J. H. Huntsman, of Tucson, is about 2½ miles north of Dos Cabezas and immediately east of the Dives group. This deposit was located in 1878 as the Juniper mine.[163] Hamilton[164] states that, during the early eighties, the Juniper claim was one of the most important in the district and, prior to 1881, produced $6,000 worth of gold and silver. In 1881-1882, 100 tons of its ore, containing $45 worth of gold per ton, was treated in a mill at Dos Cabezas. During the early nineties, the Casey brothers worked the mine in a small way. A small production was made in 1915-1917. In 1917, the owners of the property organized the Dos Cabezas Gold Ridge Mining Corporation which carried on some development work. During 1933-1934, William Dorsey operated the mine with twelve to fifteen sub-lessees. Their production from July, 1933, to June, 1934, amounted to 578 tons of ore that contained an average of 0.637 ounce of gold and 0.4 to 2.1 ounces of silver per ton and yielded a total net smelter return of $6,785.

The gold-bearing veins of this property occur within the same belt of shales as the Dives mine. (See Page 117). Underground workings are from an upper and a lower adit, connected with a winze. The upper adit has about 400 feet of drifts and stopes on a vein that ranges up to 6 feet in width. The lower adit connects with a 125-foot winze and some 700 feet of exploratory drifts on several veins.

The Gold Ridge veins consist of coarse-textured white quartz with erratically scattered bunches and disseminations of galena, pyrite, and chalcopyrite. Their gold occurs in the sulphides, mainly the galena, and is free milling only near the surface.

GOLD PRINCE MINE

The Gold Prince mine is east of the Gold Ridge property and 2¾ miles northeast of Dos Cabezas, at an altitude of 5,900 feet. It was located in 1878 as the Murphy mine. During the eighties,

---

[162] Oral communication.
[163] Unpublished notes of M. A. Allen, 1920.
[164] Hamilton, P., Resources of Arizona, 1881 and 1883.

T. C. Bain mined small amounts of high-grade ore from the property.[165] From 1918 to 1921, the Gold Prince Mining Company did more than 3,000 feet of underground development work and made a small production with a 25-ton mill. In 1931, the Dos Cabezas Gold Mining Company did some development work on the property and shipped several cars of gold ore. During 1932-1933, lessees shipped about fifty-four cars of ore that is reported to have averaged $12 in gold per ton.

Here, the westward-trending belt of metamorphosed, steeply northward-dipping Cretaceous sandstones and shales, described on page 117, forms a moderately hilly terraine. According to Leonard,[166] "The ore deposit consists of a series of lenticular bodies of gray quartz containing auriferous galena and pyrite, which make up a vein system or lode confined to an intensely sheared fracture zone in shale country rock. The shear zone is from 35 to 40 feet wide between very definite hanging and foot walls. The individual lenticular quartz veins range up to 5 feet in width and about 100 feet in length. They occur on both walls and in interconnecting and overlapping lenses in the sheared, schistose rock within the vein zone. The general strike of the vein system is about N. 70° W. and its dip 65° S.

"The surface ore was possibly somewhat richer than the ore that occurs below, owing to the removal of sulphides and substances other than gold. Oxidation and solution of sulphides does not appear to have extended for more than about 200 feet downward from the surface." The shale shows considerable alteration to graphite.

Underground workings of the Gold Prince mine include five adits with a total of more than 3,100 feet of drifts within a vertical range of 750 feet. The lowest or No. 5 tunnel, which is located near the eastern end of the property, largely in granite and diabase, trends northward for 900 feet and obliquely for 700 feet through a fault zone 500 feet wide.

LE ROY PROPERTY

The Le Roy property is 1½ miles northeast of Dos Cabezas. Its principal claims, which were located in 1878, passed through several ownerships, and were obtained by the Le Roy Consolidated Mines Company prior to 1920. A few thousand tons of gold-silver-lead ore were produced, but no records or estimates of the amount are available. In 1925-1926, the Dorsey brothers shipped, from the Le Roy shaft, five cars of carbonate ore that contained from $35 to $40 worth of gold, silver, and lead per ton.[167] In 1926-1927, the Arilead Company is reported to have shipped several cars of ore from the Climax shaft. During the following year, dump material was treated in a small mill. Dur-

165 Allen, M. A., unpublished notes, 1920.
166 Leonard, R. J., unpublished notes, 1933.
167 Oral communication from Wm. Dorsey.

ing 1928-1933, several cars of ore were shipped from the mine. In 1933, A. M. Bell installed a small mill on the property and produced some concentrates.

Here, granite, intruded by diabase dikes, forms rolling hills. The vein system strikes northeastward and dips about 65° SE. Its ore consists of coarse-textured grayish-white quartz with scattered pyrite, galena, sphalerite, and chalcopyrite.

Underground workings[168] on the Le Roy claim include an inclined shaft, more than 300 feet deep, with water at 70 feet. On the 70-foot level, the vein is 3 to 4 feet wide. Developments on the Climax claim include a 300-foot inclined shaft and more than 2,000 feet of workings. The vein ranges in width from a few to 8 inches and in places separates into a stringer lode 4 or 5 feet wide. Its ore occurs in erratically distributed bunches.

GOLDEN RULE DISTRICT

GOLDEN RULE OR OLD TERRIBLE MINE

The Golden Rule or Old Terrible mine, of northern Cochise County, is ¾ mile south of Manzoro, a siding on the Southern Pacific Railway.

This property was located during the late seventies.[169] In 1883, the Tucson Star and the U. S. Mint Report credited it with a production of $125,000 in gold. A yield of $30,000 was reported for 1884, after which the next recorded output was in 1891 when $12,000 worth of ore was shipped to Pinos Altos, New Mexico. In 1897, the mine was acquired by the Golden Queen Consolidated Gold Mining Company which built a small mill. Intermittent production continued through 1902 during which period the company was reorganized or purchased by the Old Terrible Mining Company. From 1905 to 1908, the Manzoro Gold Mining Company operated the property. No work was reported for nine years afterward. Small intermittent production, largely by lessees, has continued since 1916. In 1933, the property was owned by Mrs. E. M. Jackson, of Benson.

The recorded production from 1833 through 1929 amounts to 9,543 ounces of gold and 317,088 pounds of lead, worth about $224,000.

The mine is at the northeastern foot of the Dragoon Mountains where cherty, dolomitic, Cambrian Abrigo limestone strikes westward, dips 30° to 40° N., and is intruded by a small stock of granitic or monzonitic porphyry.

Mining has been done principally on three veins that lie from 25 to 40 feet apart, parallel to the bedding of the limestone. These veins have smooth, regular walls and are traceable for a

---

[168] Unpublished notes of Carl Lausen, 1923.
[169] History abstracted from unpublished notes of J. B. Tenney.

few hundred feet. Their filling consists of coarsely crystalline, grayish-white quartz which locally is somewhat banded to brecciated. In places it shows abundant vugs which contain hematite, limonite, calcite, cerrussite, anglesite, and galena. The gold is reported to occur mainly in the iron oxides and to a small extent in the quartz. On two of the veins, stopes about 75 to 100 feet long by 2 to 3 feet wide extend for 50 to 60 feet from the surface.

TOMBSTONE DISTRICT

The Tombstone district, in southwestern Cochise County, 20 miles northwest of Bisbee, has been noted mainly for rich silver deposits that were most actively worked during the latter part of the past century. From 1879 to 1932, inclusive, this district produced more than 29,843,800 ounces of silver, 35,669,800 pounds of lead, and $5,127,300 worth of gold, besides considerable copper, zinc, and manganese. For the last several years, some of the mines, operated largely by lessees, have yielded notable amounts of gold ore.

From November, 1933, to June, 1934, the Tombstone Development Company shipped 6,309 tons of ore that contained an average of 5.7 per cent of lead, 0.288 ounces of gold, and 15.31 ounces of silver per ton. Of this amount, 2,628 tons that averaged 8.9 per cent of lead, 0.338 ounces of gold, and 12.15 ounces of silver per ton came from the Engine mine, while 3,681 tons were mined by lessees mainly from the Head Center, Tranquility, Silver Thread, Toughnut, West Side, Flora Morrison, and Little Joe mines. Including lessees, about 160 men were employed.

Tombstone is at an altitude of 4,500 feet on a gravel-floored pediment at the northern margin of a small group of low, dissected mountains, called the Tombstone Hills, that attain an altitude of about 5,300 feet.

Early descriptions of the geology and ore deposits of the district were written by Blake and by Church.[170]

Later, Ransome[171] studied the district in detail but published only brief descriptions of its geology and ore deposits.

As interpreted by Ransome, the basement rocks are fine-grained Pinal schist with intruded gneissic granite which out-

---

[170] Blake, Wm. P., The geology and veins of Tombstone, Arizona: Am. Inst. Min. Engr., Trans., vol. 10, pp. 334-45, 1882; Tombstone and its mines: Am. Inst. Min. Eng., Trans., vol. 34, pp. 668-70, 1903.
    Church, J. A., The Tombstone, Arizona, mining district: Am. Inst. Min. Eng., Trans., vol. 33, pp. 3-37, 1902.
[171] Ransome, F. L., Deposits of manganese ore in Arizona: U. S. Geol. Survey Bull. 710, pp. 101-103, 1920; Some Paleozoic sections in Arizona and their correlation: U. S. Geol. Survey Prof. Paper 98, pp. 148-49, 1916; Darton, N. H., A résumé of Arizona geology: Univ. of Arizona, Ariz. Bureau of Mines Bull. 119, pp. 290-91, 1925.

crop in a small area south of the principal mines. The unconformably overlying Paleozoic quartzite and limestones have a total thickness of 4,000 to 5,000 feet, with 2,500 to 3,500 feet of Mississippian Escabrosa and Pennsylvanian Naco limestones as their upper formations. Unconformably overlying the Naco limestone is a Mesozoic series of conglomerate, thick-bedded quartzites, and shales, with two or three lenses of soft, bluish-gray limestone. These formations are intruded by large bodies of quartz monzonite and by dikes of quartz monzonite-porphyry and diorite-porphyry. The region is complexly faulted, and, particularly just south of Tombstone, the strata are closely folded.

The ores of the Tombstone district occur in at least three ways: (1) Irregular replacements of the blue and Naco limestones, particularly in zones of fissuring that follow the crests of anticlines, (2) in fissures, and (3) in altered porphyry dikes. Some of the replacement bodies attain very large size. Most of the ore mined came from above the water level, which is approximately 560 to 700 feet below the surface. The ores contain mainly hematite, limonite, cerussite, horn silver, and gold with locally abundant argentiferous galena, sphalerite, pyrite, alabandite, malachite, chrysocolla, psilomelane, and wulfenite. The oxidized copper minerals commonly occur in the outer margins of the ore bodies.

Gold ores: According to Church,[172] the proportion of gold to silver by weight may have been as low as 1:400 in some of the near-surface ores, but it shows a decided increase with depth. Fifty-five shipments from the deepest workings of the West Side mine yielded an average of $17.20 in gold per ton or probably four times as much as the ore from near the surface. In the Contention mine, a 140-foot drift about 90 feet below the water level gave an average assay of more than 5 ounces of gold per ton. Two shipments from below the water level of the Lucky Cuss contained 1.7 ounces of gold per ton.[173]

The oxidized silver-lead deposits that occur within approximately ½ mile south of Tombstone are locally rich in gold. As already stated on page 122, the 6,309 tons of ore shipped prior to June, 1934, by the Tombstone Development Company averaged 0.288 ounces in gold per ton. The gold ore is very erratically distributed and can generally be detected only by assays. It is commonly but not invariably siliceous. The richest material, according to Ed. Holderness, of the Tombstone Development Company, is associated with an undetermined amorphous, yellowish-green mineral. This mineral, according to microchemical tests by Robert Heineman, of the Arizona Bureau of Mines, consists mainly of lead and arsenic, together with a little zinc and copper.

---

[172] Work cited, pp. 34-35.
[173] Church, J. A., work cited.

Most of the gold of the district occurs free, but too finely divided to be revealed by panning. Some of the altered dikes, as in the Head Center mine, contain visible flakes of gold, particularly near their walls.

CHAPTER V—YUMA COUNTY

GENERAL GEOGRAPHY

Yuma County, as shown on Figure 7, comprises an area about 156 miles long by 87 miles wide. It is made up of broad desert plains, ridged with sharply eroded fault-block mountains that predominantly trend north-northwestward. Most of these ranges do not exceed 35 miles in length and, except in the central and northern portions of the area, are narrow. They attain altitudes of 2,000 to 4,000 feet in the southern and western regions, and more than 5,000 feet in the Harquahala Mountains, in the northeastern part of the county.

The area drains to the Colorado River (see Figure 7), which is the only perennial stream in this region.

The highest and lowest temperatures on record for Yuma (altitude 141 feet) are 119° and 22°, while for Salome (altitude 1,775 feet), they are 117° and 16°, respectively. The normal annual rainfall for Yuma is 3.10 inches, and for Salome 9.16 inches.

Except along the Colorado, Gila, and Williams rivers, the vegetation is of desert type.

As shown by Figure 7, the Southern Pacific and Santa Fe railways cross Yuma County. Various highways, improved roads, and desert car trails traverse the area and lead to the mining districts (see Figure 7). Because of their ruggedness and aridity, most of the mountain areas are difficult to prospect, particularly during summer.

GENERAL GEOLOGY

The mountains of Yuma County include pre-Cambrian to Mesozoic schists and gneiss; pre-Cambrian to Tertiary granites; Paleozoic to Tertiary sedimentary rocks, in part metamorphosed; and Cretaceous to Quaternary volcanic rocks. The intermont plains are floored with great thicknesses of loosely consolidated Tertiary and Quaternary sediments.

GOLD DEPOSITS

Yuma County, which ranks fourth among the gold-producing counties of Arizona has yielded about $13,250,000 worth of gold of which nearly $10,000,000 worth has come from lode gold mines. The greater part of this production was made by the Kofa, Fortuna, and Harquahala mines.

As indicated by Figure 7, the gold districts are mainly in the northern, central, and southwestern portions of the county.

With few exceptions, the deposits of economic importance have

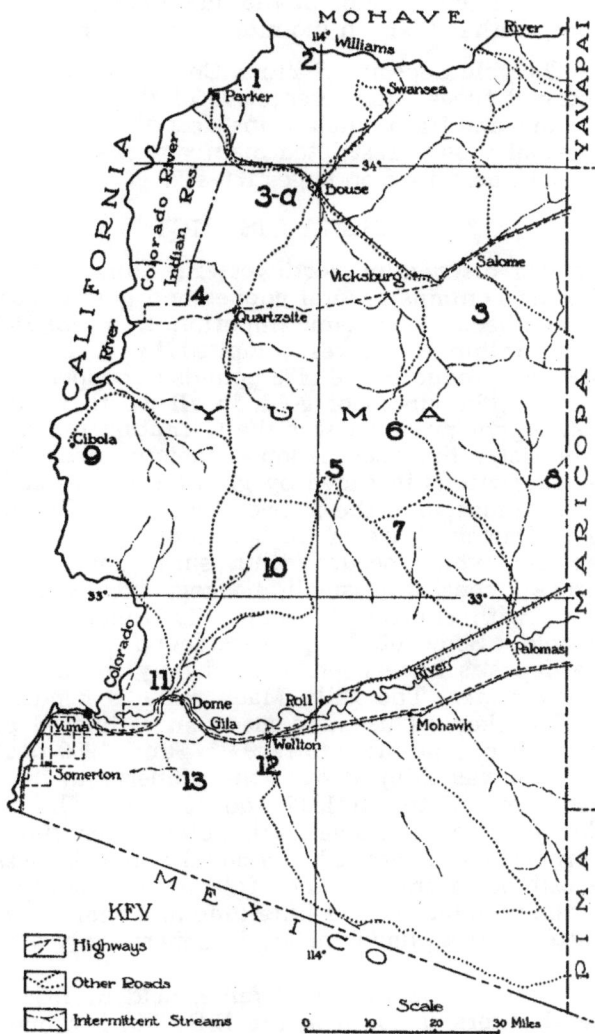

Figure 7.—Map showing location of lode gold districts in Yuma County.

1 Cienega	7 Tank
2 Planet	8 Gila Bend
3 Ellsworth	9 Trigo Mountains
3-a Plomosa	10 Castle Dome
4 La Paz	11 Las Flores
5 Kofa	12 La Posa
6 Sheep Tanks	13 La Fortuna

been found near the margins of the mountains, on pediments or gentle slopes rather than on mountain sides or high ridges.[174]

Types: The gold deposits of Yuma County are of mesothermal and epithermal types. Representative of the mesothermal type are the Fortuna and Harquahala veins, described on pages 128, 152. The epithermal type is restricted mainly to the Kofa and Sheep Tanks districts, described on pages 136-147.

CIENEGA DISTRICT

The Cienega district, in northwestern Yuma County, northeast of Parker, contains several copper and gold deposits, a few of which have been of economic importance. From 1870 to 1929, inclusive, according to figures compiled by J. B. Tenney, the Cienega district produced 1,182,473 pounds of copper, 1,935 ounces of silver, and 7,125 ounces of gold, in all worth about $402,000. The largest copper producer was the Carnation mine.

In this district, a thick succession of metamorphosed limestone, shale, and quartzite is intruded by granitic masses and overlain, at the southern margin of the district, by basalt. The limestones, according to Darton,[175] are of Carboniferous age.

The deposits, which occur within shear zones in the sedimentary rocks, consist of small, pockety bodies of brecciated country rock with chrysocolla, malachite, limonite, specularite, and flaky gold. Some of the pockets found were very rich in gold. The deposits have been worked only to shallow depths.

Billy Mack mine: The Billy Mack mine is 8 miles by road northeast of Parker. This mine made an estimated production of $65,000 worth of gold prior to 1909.[176] From 1909 to 1911, some copper-gold ore was shipped from the property to the Swansea, Arizona, smelter. Between 1917 and 1920, the Illinois-Arizona Copper Company made a small production from the mine. In 1933, the property was worked to a small extent by lessees.

This deposit, according to Bancroft,[177] occurs mainly along the contact of the metamorphosed limestone and shale. The ore occurs as scattered, irregular bunches, in general only a few inches wide.

Lion Hill mine: The Lion Hill mine, held in 1934 by W. H. Manning and others, is south of the Billy Mack property and 7 miles by road northeast of Parker.

Between 1917 and 1920, the Illinois-Arizona Copper Company shipped some copper-gold ore from this property. From 1927-1930, the Lion Hill Gold Mining Company shipped several cars of rich gold-copper ore. During 1931-1934, H. Sloan operated a 25-ton amalgamation mill on the property. The total production

---

[174] See Ariz. Bureau of Mines Bull. 134, p. 46.
[175] Darton, N. H., A résumé of Arizona geology: Univ. of Ariz., Ariz. Bureau of Mines Bull. 119, p. 218, 1925.
[176] Bancroft, H., U. S. Geol. Survey Bull. 451, pp. 74-76. 1911.
[177] U. S. Geol. Survey Bull. 451, pp. 74-76, 1911.

of the mine, according to Mr. Sloan,[178] amounts to about $30,000 in gold. The ore recently milled contained also a trace of silver and about 0.5 per cent of copper.

The deposit, which is of the type described on page 126, outcrops near the top of a low ridge of metamorphosed limestone that has been considerably faulted and jointed. Its underground workings include about 2,000 feet of drifts, together with some winzes, raises, and stopes.

Rio Vista Northside mine: The Rio Vista Northside property of twelve claims, held by E. S. Osborne and J. E. Ransford, is west of the Lion Hill mine and 5 miles by road northeast of Parker.

In 1918-1919, this property yielded three cars of sorted, highgrade gold-copper ore.[179] During recent years, it has been developed by a 224-foot incline and a 275-foot tunnel.

The deposit occurs in a low ridge, within a wide, silicified, brecciated zone that strikes S. 15° W. and separates metamorphosed limestone on the west from schistose quartzite on the east. This zone, as revealed by shallow surface cuts, contains numerous lenticular bodies of the type described on page 126.

Capilano mine: The Capilano property, owned by C. A. Botzun, is north of the Rio Vista, at the end of a low ridge. This mine has made a small production of rich gold-copper ore. The deposit, which is similar to that of the Rio Vista mine, occurs within a gently eastward-dipping brecciated zone. It has been explored by a shallow shaft and several surface cuts.

Sue mine: The Sue mine, about ½ mile north of the Rio Vista property, is reported to have been worked intermittently since the sixties and to have produced some rich ore.

This deposit, which is of the general type outlined on page 126, occurs within a brecciated zone that strikes N. 25° W., and dips about 80° SW. It has been opened by a shaft about 100 feet deep. The country rock is schistose quartzite and siliceous limestone.

PLANET DISTRICT

The Planet copper mining district, in the vicinity of the Williams River, north of Bouse, has produced a little gold.

This rugged dissected region consists of granitic gneiss and schist, overlain by metamorphosed limestones, amphibolites, and mica schists. Many of its geologic features have been outlined by Bancroft[180] and by Darton.[180]

---

[178] Written communication.

[179] Oral communication from Mr. Osborne.

[180] Bancroft, H., Reconnaissance of the ore deposits in northern Yuma County, Arizona, U. S. Geol. Survey Bull. 451, pp. 46-59, 1911.

[180] Darton, N. H., A résumé of Arizona geology: Univ. of Ariz., Ariz. Bureau of Mines Bull. 119, pp. 215-18, 1925.

PLANET LEASE

During 1933-1934, M. W. Martinet, R. W. Geitlinger, D. M. Wenceslaw, and E. L. Craig, lessees, mined more than a carload of rich gold ore from a newly discovered deposit on the Planet Copper Mining Company's ground, about 2 miles south of the Williams River and 28 miles by road north of Bouse.

Here, the prevailing rock is bleached siliceous schist which locally contains stringers of quartz and films of iron and copper oxides.

The deposit occurs within a brecciated zone, several feet wide, that strikes northeastward and dips .about 30° SE. The ore occurs as irregular streaks and small, discontinuous pockets of brecciated to pulverized sugary quartz with abundant dark-red to black iron oxide and sericite. Upon panning, this material reveals abundant finely divided particles of gold.

In January, 1934, workings on the deposit consisted of a shallow shaft and a short tunnel.

HARQUAHALA MOUNTAINS (ELLSWORTH DISTRICT)

BONANZA (HARQUAHALA) AND GOLDEN EAGLE MINES

Situation: The Bonanza or Harquahala mine is in the southwestern portion of the Harquahala Mountains, at an elevation of about 1,800 feet above sea level. It is accessible from the railway at Salome by 9 miles of road. The Golden Eagle mine is about a mile northeast of the Bonanza. These mines are held by the Bonanza and Golden Eagle Mining Company.

History: The Bonanza and Golden Eagle veins were discovered[181] in 1888 and sold to Hubbard and Bowers who organized the Bonanza Mining Company. It is reported that a clean-up worth $36,000 was made from a week's run of a small amalgamation mill. A 20-stamp amalgamation mill, erected in 1891, made an estimated production of $1,600,000 in bullion within three years.

In 1893, The Harqua Hala Gold Mining Company, Ltd., a British syndicate, purchased the property for $1,250,000, remodeled the mill, and sank a new shaft. During 1895, a 150-ton cyanide plant was built to treat the accumulated tailings which ran from $3 to $5 per ton. Both the ore body and the tailings dump were exhausted by the end of 1897, and the mine was sold back to Mr. Hubbard in 1899. The total production by the British company amounted to $750,000 in bullion, of which about $125,000 was profit.

After a few months' operation, the mines remained idle until 1906 when the Harqua Hala Mining Company was organized.

---

[181] Abstracted from unpublished notes of J. B. Tenney.

PRODUCTION BONANZA (HARQUAHALA) AND GOLDEN EAGLE MINES.

(Data compiled by J. B. Tenney).

Year	Price of lead	Lbs. lead	Ounces gold	Gross value	Remarks
1891-1893			77,407	$1,600,000	Hubbard and Bowers (est.)
1893			6,410	132,500	Harqua Hala Gold Mining Company, Ltd.
1894			13,148	271,775	Harqua Hala Gold Mining Company, Ltd.
1895			6,803	140,615	Harqua Hala Gold Mining Company, Ltd.
1896			4,053	83,778	Harqua Hala Gold Mining Company and lessees.
1897			5,721	118,262	Harqua Hala Gold Mining Company and lessees.
1898			149	3,070	Harqua Hala Gold Mining Company and lessees.
1899			484	10,000	**Hubbard (est.)**
1907			1,449	29,960	Harqua Hala Mining Company
1908			1,113	23,000	Harqua Hala Mining Company
1914			919	19,000	Yuma Warrior Mining Company
1916			1,451	30,000	Yuma Warrior Mining Company
1922-1923			726	15,000	Lessees.
1925	.087	55,000	73	6,285	Lessees on Golden Eagle.
1926	.080	53,600	73	5,788	Lessees on Golden Eagle.
1928	.058	14,600	242	13,468	Lessees on Golden Eagle.
1929	.063	19,000	339	8,197	Lessees on Golden Eagle.
Total 1891—1929		142,200	120,560	$2,510,698	

By the end of 1908, this company produced about $53,000 in gold bullion, at little or no profit.

From 1913 to 1916, the Yuma Warrior Mining Company produced $30,000 from the mines and $19,000 from tailings.

A small production, chiefly by lessees, was made from 1922 to 1933, as shown in part on page 129. The total production from the Bonanza and Golden Eagle mines amounts to about $2,500,000.

Early in 1934, the Bonanza mine was under lease to the Harquahala Gold Mines Company, but no underground work was in progress. W. L. Hart and associates treated about 1,000 tons of the old Harquahala mill tailings by leaching and cyanidation, but suspended operations in April, 1934. Lessees made a small production from the Golden Eagle mine.

Topography and geology: This portion of the range consists of small, sharply eroded mountains flanked by desert plains. Bancroft[182] states that the following succession of rocks is present: Coarse-grained basal granite, exposed north of the Golden Eagle mine; quartzitic grits; limestone and shale; thin conglomerate; and shale and limestone with some dolomite and conglomerate. These rocks are more or less metamorphosed and complexly faulted. Darton[183] has shown that the limestones, in part at least, are of Carboniferous age.

Bonanza or Harquahala deposit:[184] The Bonanza or Harquahala mine is at the northwestern base of Martin Peak, in the southwestern portion of the Harquahala Mountains. The deposit occurs within a zone of faulting that strikes northward and extends through the limestone, shale, and quartzite into the basal granite. Its main shear zone dips 45° W. and is joined by a lesser fault that dips 45° E. The ore shoots that were mined in the "Castle Garden" stope occurred within these two shear zones and ranged from a few inches to many feet in width. As indicated by Bancroft's map, the several stopes of the mine occupied an A-shaped area about 500 feet long by some 450 feet wide on the south, all above the fifth or water level, which is 170 feet from the surface. In this part of the mine, the gangue consists of soft red hematite, with quartz, calcite, brecciated country rock, and a little gypsum. In places, large masses of gold occurred intimately associated with quartz. Very little silver was present. Below water level, the ore is pyritic. The granite, which appears on the sixth and seventh levels, shows intense sericitization.

Workings of the Bonanza or Harquahala mine include an inclined shaft and many hundred feet of drifts on seven levels. The stopes are all on the first five levels.

---

[182] Bancroft, H., Reconnaissance of the ore deposits in northern Yuma County, Arizona: U. S. Geol. Survey Bull. 451, pp. 106-109, 1911.

[183] Darton, N. H., A résumé of Arizona geology: Univ. of Ariz., Ariz. Bureau of Mines Bull. 119, pp. 221-23.

[184] Abstracted from Bancroft, work cited.

The Harquahala tailings dump is reported by Miles Carpenter[185] to contain about 29,000 tons that average 0.124 ounces of gold and 0.40 ounces of silver per ton.

Golden Eagle Deposit: The Golden Eagle mine is about a mile northeast of Harquahala, at the northern base of a low ridge. The main vein strikes S. 20° W., dips about 50° SE., and occurs in quartzite. A few feet farther northeast is a parallel vein that dips about 85° NW. The gangue consists largely of coarse-textured, grayish-white quartz. Iron oxide is abundant above the water or the 300-foot level on the incline, below which pyrite, chalcopyrite, and galena occur.

Workings of the Golden Eagle mine include a 400-foot shaft inclined at 45°, with about 450 feet of drifts on the lower level and stopes that extend from the surface to the 300-foot level.

As indicated by the stopes, the ore shoots on both veins were numerous and pockety. Some of them apparently were more than 15 feet wide. Bancroft[186] states that two samples of sulphide ore from the lower levels contained 0.25 and 4.84 per cent of copper, 1.32 and 2.88 ounces of silver, and 0.48 and 1.12 ounces of gold per ton.

SOCORRO MINE

The Socorro mine, at the southern base of the Harquahala Mountains, is accessible from Salome by 11 miles of road.

The Socorro Gold Mining Company[187] acquired this mine in 1901 and, within four years, sank a 375-foot shaft and ran 2,300 feet of drifts. A 20-stamp mill, equipped for amalgamation, concentration and cyanidation was built in 1904. The milling process used has been described by Smith.[188] Intermitent operations, carried on from 1906 to 1914, yielded about $20,000 in gold bullion.

In this vicinity, sheared, coarse-grained granite is overlain by about 150 feet of quartzite, followed by several hundred feet of metamorphosed, yellowish-brown limestone and shale of Carboniferous age. These rocks have been considerably faulted, and the strata dip 30° to 80° eastward.

The Socorro vein, as described by Bancroft,[189] occupies a fault fissure within the granite and sedimentary rocks. It strikes westward, dips 26° N., and ranges in width from a few inches to several feet. Free-milling ore, consisting of white quartz and oxidized gold-bearing iron minerals, was mined above the 250-foot level. Below this level, pyrite is scattered through the quartz

---

[185] Oral communication.
[186] Work cited, p. 109.
[187] History abstracted from unpublished notes of J. B. Tenney.
[188] Smith, F. C., The cyanidation of raw pyritic concentrates: Am. Inst. Min. Eng., Trans., vol. 37, pp. 570-75, 1907.
[189] Bancroft, H., Ore deposits of northern Yuma County, Arizona: U. S. Geol. Survey Bull 451, p. 112, 1911.

and is locally abundant in the wall rock. The vein material shows considerable brecciation and recementation by silica.

At the time of Bancroft's visit (1909), water stood near the 250-foot level of the shaft or 110 feet below the surface. The shaft inclines 26° N.

SAN MARCOS MINE

The San Marcos mine, at the northern base of the Harquahala Mountains, is 5 miles by road southeast of Wenden.

According to local reports, this deposit was located about 1897. Later, it was acquired by the Pittsburgh Harqua Hala Gold Mining Company which erected camp buildings and installed machinery but gave up the enterprise after sinking an inclined shaft to a depth of 540 feet.

Bancroft[190] states that the total production up to October, 1909, was worth approximately $12,400, most of which was made prior to 1906. About 1915, the ground was relocated by Pete Smith and associates who, for a short time, carried on small-scale operations. No production was reported since 1919.

Here, sheared granite has been intruded by aplitic and basic dikes and considerably faulted. The deposit occurs within a shear zone that strikes northeastward and dips about 30° NW.[191] The vein filling consists principally of quartz, brecciated country rock, and abundant iron oxide. Above the first level, it was from 18 to 24 inches wide, and contained high-grade gold ore. Near this level, the vein, because of either pinching out or flattening, disappears in the roof. At greater depth, only small lenses of the gold-bearing quartz were found. Oxidation extends to the bottom of the mine.

In 1909, there were about 1,200 feet of drifts on the four levels of the incline.

HERCULES MINE[192]

The Hercules mine, at the northern base of the Harquahala Mountains, is 5 miles by road southeast of Salome. Its production prior to 1909 is reported to have been about $10,000 worth of gold. In 1934, the property was held by the Hercules Gold Company.

The deposit consists of discontinuous, lenticular quartz veins in gneissic quartz diorite, intruded by dioritic dikes. It has been opened by two shallow shafts that incline about 60° N. The vein exposed in the eastern shaft strikes S. 75° W., dips 60° N., and is from 8 to 24 inches wide. It shows brecciation and contains abundant limonite.

---

190 Work cited, p. 113.
191 Description abstracted from Bancroft, work cited, pp. 113-14.
192 Abstracted from Bancroft, work cited, pp. 109-10.

HIDDEN TREASURE MINE

The Hidden Treasure property, held in 1934 by W. Neal and G. Meyers, consists of three claims near the southern base of the Harquahala Mountains, in northeastern Yuma County. It is accessible by 5 miles of road that branches eastward from the Salome-Hassayampa road at a point 11 miles from Salome.

This deposit was located in 1932 by G. Myers and J. Lazure. During that year, according to J. V. Allison,[193] it yielded twenty-two cars of shipping ore which averaged $21.55 in gold per ton, and four cars which ranged from $6 to $12 per ton. Only a little work was done on the property during 1933.

This portion of the Harquahala Range is made up mostly of complexly faulted, metamorphosed limestones of upper Carboniferous age. The ore deposit, which outcrops on a steep slope about 300 feet above the plain, is a tabular replacement of a southwestward trending, steeply dipping strike-fault zone. It has been explored by a 200-foot tunnel that connects with a stope about 140 feet long by 8 or 9 feet high and up to 20 feet wide. The ore occurs as irregular cellular masses of limonite and calcite together with more or less quartzite and silicified limestone. In places, chrysocolla and manganese dioxide are abundant. The gold occurs as fine to mediumly coarse particles, mainly with the limonite. The wall rock, which shows intense silicification with some sericitization, in places is quartzitic.

ALASKAN MINE

The Alaskan mine, on the plain south of Harquahala Mountain, in northeastern Yuma County, is accessible by 8 miles of a road that branches eastward from the Salome-Hassayampa road at a point 13 miles from Salome and continues to Aguila.

This deposit was discovered in 1920 by A. Johnson, the present owner, and has been worked intermittently by several different concerns. The total production of the mine amounts to approximately 1,200 tons of ore which contained from $6 to $16 worth of gold per ton. From January, 1925, to May, 1933, the mine produced 722 tons of ore that averaged 1.69 per cent of copper, 0.503 ounces of gold, and 0.37 ounces of silver per ton. The Alaskan Mines, Inc., treated approximately 800[194] tons in a small flotation plant, but the results were unsatisfactory. Water for all operations was hauled from distant places.

Here, a gravel-mantled pediment is trenched by arroyos which, in a few places, expose steeply dipping metamorphosed, impure shales of probable Mesozoic age, intruded by dikes of altered, basic porphyry.

When visited in February, 1934, workings on the Alaskan prop-

---

[193] Oral communication.
[194] Oral communication from A. Johnson.

erty included a 187-foot inclined shaft that connects on the west with a few hundred feet of shallow openings. The shaft inclines about 25° southward and passes beneath the ore deposit, which occurs within a gently southward-dipping brecciated zone that, due to flattening southward from the collar of the shaft, is only a few feet below the surface. In places, the hanging wall has been eroded away, and the breccia covered by surface gravels. The ore consists of brecciated, silicified shale with small fragments of coarsely crystalline, shiny gray quartz, cemented with limonite, chrysocolla, calcite, and minor amounts of fluorite. Yellow lead oxide stain is sparingly present, and a little manganese dioxide occurs in places. According to Mr. Johnson, the gold, which is rather finely divided, occurs mainly where the chrysocolla and reddish limonite are relatively abundant.

The breccia appears to represent a thrust fault zone into which the ore-bearing solutions entered from below along transverse fractures. On the south, the ore body is cut off by a steeply northward-dipping fault, but the structural conditions that might determine possible continuation of the ore westward have not been determined.

PLOMOSA DISTRICT

DUTCHMAN MINE

The Dutchman mine, about 6 miles by road northwest of Bouse, is reported to have been located in the early days by Bouse and McMann. In 1912-1913, according to George Y. Lee,[195] caretaker of the property, it produced more than $20,000 worth of ore that averaged $30 per ton in gold and copper. In 1914, a small amalgamation-concentration mill was built near Bouse Wash, 4 miles southeast of the mine, where water was obtained at a depth of 88 feet. After a short run, however, this mill proved to be unsatisfactory. Some underground development work was done in 1926-1928 by the Great Western Gold and Copper Mining and Milling Company, and in 1929 by the Protection Gold and Copper Company which produced a carload of ore. Intermittent, small-scale operations were carried on at the mine until 1933 when the property was sold at Sheriff's sale and the surface equipment dismantled. In 1934, it was held by F. G. Bradbury.

The Dutchman mine is in the low, narrow northern portion of the Plomosa Mountains which here are composed of metamorphosed limestone, probable Cretaceous shales, and intrusive granite. The deposit occurs within a steeply dipping fault zone in metamorphosed red hematitic shales. According to Mr. Lee, the mine workings include a 400-foot shaft, now filled to the 150-foot level, about 170 feet of drifts on the 100-foot level, and some stopes near the surface and on the 100-foot level. The gold oc-

---

[195] Oral communication.

curs as small to visible flakes in hematitic, locally copper-stained shale.

BLUE SLATE MINE

The Blue Slate mine, about ½ mile east of the Dutchman mine, is reported to have been located at an early date by Bouse and McMann who conducted profitable arrastring operations on the near-surface ore. Intermittent, small-scale operations which resulted in a small production were carried on from about 1918 to 1929. In 1929, the mine was operated by the Protection Gold and Copper Company which shipped a little ore. In 1934, the property was held by the Great Western Gold and Copper Mining Company.

The vein, which is about 2 feet wide, occurs within a fault zone that strikes about N. 20° W., dips steeply northeastward, and cuts metamorphosed, sandy reddish shales. According to George Y. Lee, it was opened by two shafts, 150 and 60 feet deep, with very little drifting.[196] Most of the ore mined was from surface cuts on the vein. As seen in these cuts, the gangue is coarse-grained, vuggy, grayish-white quartz with more or less limonite and hematite. The vein walls show much hematite and some copper stain.

OLD MAID MINE

The Old Maid mine is in the northern portion of the Plomosa Mountains, 5 miles by road northwest of Bouse. This mine, which was opened by shallow workings several years ago, was obtained by the Old Maid Mining Company in 1924. This company sank a 260-foot vertical shaft and ran a total of nearly 1,000 feet of drifts on five levels. In 1928, according to the U. S. Mineral Resources, a car of ore that averaged nearly 2 ounces of gold per ton was shipped. During the past few years, only small-scale, intermittent operations have been carried on.

The deposit occurs in metamorphosed sedimentary rocks of probable Cretaceous age that are intruded by granitic porphyry. The vein strikes northward and dips 20° to 60° W. It has been complexly faulted and has not been found below the 100-foot level. As seen in the upper workings, its thickness ranges from a few inches up to about 3 feet and averages about one foot. As a rule, the narrower portions are the richest. The ore consists of coarse grayish-white vuggy quartz with abundant reddish to black hematite. Its gold occurs in a finely divided condition, mainly with the hematite.

LA PAZ OR WEAVER DISTRICT

La Paz or Weaver district is in the Dome Rock Mountains, of northwestern Yuma County. The northwestern part of the dis-

---

[196] Oral communication.

trict, north of the middle of T. 4 N., is within the Colorado River Indian Reservation.

Gold-bearing quartz veins were discovered in this district during the sixties. Small-scale, intermittent work upon them has yielded a total of probably less than $100,000 worth of gold.

This portion of the Dome Rock Mountains is made up of sedimentary and igneous schists, locally intruded by granite. The larger and more persistent veins are generally parallel to the schistosity, and small gash veins cut across the schistosity.[197]

Goodman mine:[197] The Goodman mine is about 8 miles west of Quartzsite and a few miles north of U. S. Highway 60.

This mine, acording to Jones, produced about $40,000 worth of gold prior to 1900, and $9,000 more from 1900 to 1914. The ore was hauled 15 miles to Quartzsite and there treated in a small amalgamation mill. In 1934, the property was taken over by G. W. McMillen who reopened the mine and installed a small amalgamation mill on the property.

The Goodman vein strikes east-southeastward, dips from 30° to almost 90° N., and occupies a shear zone that is treaceable for more than 2 miles across the range, between La Paz placers on the west and the Middle Camp placers on the east. It occurs in quartz-epidote schist. In width, the vein ranges from less than an inch up to 40 feet and averages about 10 feet. Its filling consists of massive quartz with numerous cavities. In the oxidized zone, these cavities are more or less filled with iron oxide that contains visible free gold. Where oxidation has not been complete, gold-bearing pyrite is relatively abundant, particularly near the walls of the vein.

Underground workings include several inclined shafts and connecting tunnels. One of the inclines is 120 feet deep and connects with 260 feet of workings. One of the adit tunnels is 240 feet long.

Other properties: Jones[198] describes the Don Welsh and Mammoth prospects, on the Colorado Indian Reservation, the Golden Hope prospect, on the eastern portion of the Goodman vein, and the Mariquita prospect, north of the Oro Fino placers. So far as known, these prospects have made little or no production.

KOFA DISTRICT

Situation and accessibility: The Kofa district, at the southwestern margin of the S. H. or Kofa Mountains, is accessible from Yuma by a semi-improved road that branches eastward from the Quartzsite highway at the Stone Cabin. Unimproved roads lead from the Southern Pacific Railway at Wellton and Mohawk to

---

[197] Jones, E. L. Jr., Gold deposits near Quartzsite, Ariz.: U. S. Geol. Survey Bull. 620, p. 55, 1915.
[198] Work cited, pp. 55-57.

the district. The nearest railway point, Growler, on the Phoenix loop of the Southern Pacific, is about 35 miles distant.

Topography and geology: The bulk of the S. H. Mountains consists of a gently northward-dipping block of Tertiary rhyolites, andesites, and tuffs, capped by basalts. This mass rises to more than 4,000 feet above sea level, or 2,500 feet above the adjacent plains. Portions of its crest represent a sloping mesa, modified in many places by pinnacles and deep, groove-like canyon systems. Cliff-like slopes are rather common. The western and southern fronts are wall-like, but the northern side is very extensively dissected.

In general, only narrow pediments are exposed in the Kofa district. In places, particularly along the southern base of the range, are minor areas of schist and granite and a large area of probable Cretaceous shales, intruded by dikes of diorite, pegmatite, and monzonite-porphyry.

Extensive faulting and fracturing have affected all of the rocks, particularly those of pre-lava age.

GOLD-BEARING LODES

The gold-bearing lodes of the Kofa district are of the epithermal type but differ widely in detail. Thus, the King of Arizona lode consists of brecciated andesite-porphyry traversed by stringers of quartz and manganiferous calcite, whereas the North Star vein is locally brecciated and cemented by banded chalcedonic quartz accompanied by adularia and finely divided pyrite. The King of Arizona lode occurs in andesite, but the North Star lode is between a hanging wall of andesite and a footwall of slate.

KING OF ARIZONA MINE

Situation, history and production: The King of Arizona vein outcrops at an elevation of about 1700 feet above sea level, near the foot of a blocky spur at the southern margin of the S. H. Mountains.

This vein was discovered during the winter of 1896 by Chas. E. Eichelberger, who, with H. B. Gleason and Epes Randolph, shortly afterwards organized the King of Arizona Mining Company.[199] In June, 1897, a 5-stamp amalgamation mill was completed near Mohawk, on the Gila River, 35 miles south of the mine. In 1898, a cyanide plant for treating the tailings was added. This first mill, to which the cost of delivering ore amounted to $8 per ton, was operated until January, 1899. During the latter part of 1897, after several unsuccessful attempts, an adequate water supply was located at a depth of 1000 feet in a well 5 miles south of the property. Water was also struck at a depth of 465 feet on a well 17 miles south of the mine. In 1899, the mine was temporarily closed while a 100-ton cyanide plant was being built at the mine.

---

[199] Historical data compiled by J. B. Tenney.

Later, the mill was enlarged to a capacity of 250 tons, and about 125 men were employed on the property.

From August, 1899, to July, 1910, the mine and mill were run continuously with an average production of 200 tons a day. The total production as, shown in page 142, amounted to $3,500,000.

In 1933, the Rob Roy Development Company put down a well in the gulch about ¾ of a mile north of the mine, started construction of a 50-ton mill, and did a little development work. In 1934, the property was held by the Kofa Mining Syndicate.

Mineral deposit: Blake[200], in 1898, wrote the following notes regarding the King of Arizona vein: "This lode or vein has three well-marked divisions or layers. On the hanging wall there is a soft layer from 3 to 3½ inches wide, which averages about $800 per ton in value. Next below this there is a middle layer or ore body of quartz about 20 inches thick, which will average $190 to $200 per ton in value. The remainder of the vein, so far as it is exposed by the shaft, averages about $24 per ton. Test holes have been drilled 3 feet deep into the foot wall, and all are in ore . . . At a point about 30 feet west of the shaft and on a level with the collar of the shaft the vein has been crosscut from wall to wall and is shown to be 18 feet wide. The ore in this crosscut has about the same grade as that at the collar of the shaft. The value of the vein at the shaft has been estimated at $3,500 per lineal foot, longitudinally."

Jones,[201] who visited this property in 1914, states: "At the time of the writer's visit the mine had been closed for four years and the workings were in part inaccessible and generally in a bad state of repair below the 100-foot level . . . The mineralized zone or lode trends between N. 60° W. and W. and dips at an angle of 60° S. This zone can not be traced beyond the limits of the detached hill area. Its identity is lost a few hundred feet east of the mine shaft, but west of the shaft the vein is covered by two claims of the King of Arizona group, and it probably extends for a considerable distance beyond them. On the King of Arizona claims the vein is stoped out to the surface for 1,500 feet along its strike. The stoped areas are from a few feet to 30 feet wide and the ore body is said to average 12 feet in width. There is no mine dump, as all the material was run through the mill. The footwall of the vein is generally a well-defined slickensided plane, but the hanging wall is more indefinite. The ore body contained many small fissures and small slip planes, and most of them are parallel to the trend of the ore body, but several lie at angles with the vein, generally coming in from the hanging-wall side, and make horses of barren material. About 200 feet east of

[200] Blake, W. P., Report of the Terr. Geol., in rep't of Gov. of Ariz., 1898: Misc. Rep'ts., Washington, 1898, pp. 255-56.
[201] Jones, E. L. Jr., A reconnaisance in the Kofa Mountains, Arizona; U. S. Geol. Survey Bull. 620, pp. 157-58, 1915.

the shaft, strong cross fissures filled with calcite apparently limit the ore, for development has not proceeded beyond this point.

"The lode matter is a brecciated, generally brown to maroon andesite-porphyry. The andesite is partly silicified, particularly where the fissuring is closely spaced. Stringers of quartz and calcite traverse the lode in all directions. They vary from those of knife-blade thickness to those several feet thick. The small veinlets are composed of quartz crystals, but in those one inch or more thick the walls are commonly lined with small quartz crystals and calcite occupies the middle. The calcite is brown to black in color and is highly manganiferous. The lode matter is stained with iron and manganese oxides. The gold is said to occur free but in a very finely divided state. None was noted in the specimens collected from the vein and dump. The disintegration of the lode has produced no placer deposits.

"The ore is valuable chiefly for its gold, but it also contains silver, the two metals being present in the ratio of approximately 58 to 1 in value. The ore at the surface was very rich, and many tons of it valued at $2,000 a ton were mined. The average tenor was $40 a ton. The metal content of the ore body steadily decreased with increasing depth until at the deepest workings, 750 feet below the surface, the gold and silver content averaged less than $3 a ton. At this figure the ore could not be treated profitably and the mine was closed. The walls of the ore body diverge with increasing depth, and Mr. Ives states that at the bottom of the shaft they are 80 feet apart.

Workings: "The mine is developed by an inclined shaft 750 feet deep, drifts at the 100-foot level, and an adit at the level of the collar of the shaft. The shaft is driven on the hanging-wall side a short distance south of the outcrop of the vein and approximately 100 feet lower. The drifts extend east and west and follow the vein; some of those to the west are over 2,000 feet long, but the drifts east of the shaft are not longer than 200 feet."

Mill: The original 100-ton mill used at the King of Arizona mine has been described in the Engineering and Mining Journal, vol. 68, p. 566, 1899. A description of the later 250-ton plant was given by Jones[202] and reprinted in Arizona Bureau of Mines Bulletin 134.

Costs: Development, mining, milling, taxes, and general expenses during the last few years of operation are reported to have cost about $2.80 per ton.

NORTH STAR MINE

Situation and history: The North Star mine is at the foot of the precipitous southern wall of the main mountain mass, about

---

[202] U. S. Geol. Survey Bull. 620, p. 158.

1¾ miles north of the King of Arizona mine. A gravel-floored reentrant separates it from the spur that contains the King of Arizona lode.

The North Star vein was discovered in 1906 by Felix Mayhew, who sold it in 1907 to the Golden Star Mining Company for $350,-000.[203] In 1908, this company erected a 50-ton cyanide plant at the mine, and later doubled its capacity. Water for this plant was obtained from the King of Arizona wells. Operations continued until August, 1911, when the ore reached an unprofitable grade.

As shown on page 142, the production from this property amounted to more than $1,100,000. In addition, a considerable amount of rich ore was stolen. In 1934, the property was held by the Gold Star Mining Company.

Mineral deposit: This mine has been closed since 1911, but Jones[204] has given the following description:

"The ore body of the North Star mine is in a lode or vein of silicified andesite breccia and quartz which strikes east and dips about 60° N. The lode is 10 feet in average width at the North Star. It crops out several feet above the country rock, and from the mine it can plainly be seen extending for a considerable distance to the east and west. At the mine the hanging wall is a pink flow-banded biotite andesite, and the footwall a dark calcareous shale or slate. The shale in places contains finely disseminated pyrite. The lode probably occurs along a fault.

"The vein matter is of striking apparance. Near the surface angular fragments of the pink andesite, in places altered to green and gray tints, are cemented to an extremely hard rock by banded chalcedonic quartz, the bands being well shown by the deposition of minute crystals of sulphides, most if not all of which are pyrite. Pyrite also occurs disseminated sparingly through the altered andesite. Small vugs in this material are lined with sparkling quartz crystals. Under the microscope the chalcedonic quartz is seen to be accompanied by adularia in variable amounts. A green micaceous mineral is developed in the altered andesite, as well as a little chlorite and epidote.

"The ore is valuable chiefly for its gold content, but it also contains small amounts of silver in about the ratio of its occurence in the King of Arizona ore. The gold is said to occur free and very finely divided, associated with the fine sulphides in the chalcedonic quartz.

"As interpreted from the assay chart of the mine, the high-grade ore bodies occur in shoots or chimneys which pitch to the east. They are of variable width, and the gold content deteri-

---

[203] Oral communication from Mr. Mayhew.
[204] Work cited, pp. 159-60.

orates rapidly with increasing depth until at the fifth level the average tenor of the ore is below that required for its profitable treatment—$14 a ton. Assays in the drifts beyond the enriched parts of the lode show gold and silver in variable amounts, with a probable average value of $2 a ton. The surface ore of the North Star mine was of exceptionally high grade. One streak of ore on the footwall was said to have been worth from $6 to $20 a pound, and ore to the value of thousands of dollars was stolen. The stoped-out parts of the ore bodies averaged 10 feet in width. No. 2 shaft was sunk on the lode in the approximate center of a shoot of high-grade ore over 500 feet long, the tenor of which exceeded $50 a ton. On the second level the high-grade ore occurs in several small shoots separated by ledge matter of relatively low grade. On the third, fourth, and fifth levels these shoots appear to have joined to form a shoot which is comparable in length to that of the first level, but which shows a rapidly decreasing metal content.

"The ore of the North Star mine differs markedly from that of the King of Arizona mine in the absence of calcite and in the abundance of chalcedonic quartz and pyrite, factors which make it far less amenable to cyanidation. Several processes are necessary in order to reduce the ore to sufficient fineness to release the gold content. The cost of mining and milling is said to be $14 a ton. The disintegration of the North Star lode has produced no placer deposits.

Workings: "The mine is developed by two inclined shafts, No. 1 and No. 2, 90 and 500 feet deep, respectively, and by drifts and crosscuts on the vein at each 100-foot level. An adit from the footwall side connects with the first level from No. 2 shaft. The total length of development work is about 3,500 feet."

ECONOMIC POSSIBILITIES OF KOFA DISTRICT

In the King of Arizona and North Star mines, consistently lower-grade material is likely to occur at greater depths. These mines, however, contain large tonnages of material which was not of economic grade in 1911, but which might now be mined at a profit if the price of gold remains considerably higher than $20.67 per ounce.

Intensive search for possible undiscovered veins in this district began after the discovery of the King of Arizona lode in 1896, but the nearby North Star vein remained unlocated until 1906. Its discoverer, Felix Mahew, states[205] that all the prospectors had ignored the outcrop because it did not resemble the King of Arizona lode. Although prospecting here became more thorough after 1906, the marginal portions of the range, particularly where

---

[205] Oral communication.

PRODUCTION, KOFA DISTRICT.
(Data Compiled by J. B. Tenney)

Date	Price Silver	Ounces Silver	Ounces Gold	Total Value	Remarks:
1897-1906	.59	66,382	132,764	$2,784,063.	King of Arizona.
1907	.66	4,774 232	12,022 464	251,622 9,774	King of Arizona. Golden Star Min. & Mill. Co. (North Star).
1908	.53	6,177	16,151	337,115	King of Arizona. Golden Star Min. & Mill. Co. (North Star).
1909	.52	4,700 9,937	11,735 24,491	245,017 511,305	King of Arizona. Golden Star Min. & Mill. Co. (North Star).
1910	.54	7,492	15,945	333,590	Golden Star Min. & Mill. Co. (North Star).
1911	.53	2,731	6,099	127,514	Golden Star Min. & Mill. Co. (North Star)—(Closed August).
TOTAL 1896-1911		102,425	219,671	$4,600,000	Grand total from Jones Report, Bull. 620, U.S.G.S.
1912	.615	4,523	9,046	18,996	Lessees and small mines.
1913	.604	54	108	2,274	Placers and small mines.
1914	.553	48	96	2,000	$709 placers. Balance bullion from Independence Mine.
1915	.507	177	354	7,433	Placers and Quartette Mine.
1916	.658	166	333	7,000	Lessees on North Star, Quartette, Independence and placers (est.)
1917	.824	47	95	2,000	Lessees on North Star (est.)
1918	1.000	48	95	2,000	Placers and Lessees on North Star (est.)
1924	.67	23	47	1,000	Ironwood Mine (est.)
1925	.694	48	95	2,000	Lessee on North Star (est.)
1928	.585	12	24	500	Placers (est.)
TOTAL		107,571	229,964	$4,645,203	

pediments or rock floors are concealed by talus, merit further search.

GOLD PROSPECTS IN THE ALAMO REGION

The Alamo region, as here considered, embraces a small mineralized district in the vicinity of Alamo Spring, Cemitosa Tanks, Red Raven Wash, and Ocotillo, in the northern portion of the S. H. Mountains.

Alamo Spring, which is approximately 30 miles from Vicksburg, is accessible by about 13 miles of unimproved road that branches westward from the Sheep Tanks road at the eastern entrance of New Water Pass. The Cemitosa Tanks, which are 4½ miles northeast of Alamo Spring, are accessible by about 12 miles of desert road that branches westward from the Sheep Tanks road at a point some 8 miles south of Sheep Tanks. Red Raven Wash is about 2 miles by road southwest of Alamo Spring. The road leads westward beside this wash for 2 miles, to Ocotillo.

Many claims were located in this region about 25 years ago and have been considerably prospected from time to time. So far as known, they have produced no ore.

Here, the mountains are somewhat less cliffy and rugged than in the vicinity of Kofa. Sharp ridges and flat-topped mesas prevail, but they are separated by canyons that are rather broad, especially in the tuffaceous rocks.

The gold prospects of the Alamo region have been described by Jones.[206] The veins occur within steeply dipping brecciated zones in andesite. They range in width from a few feet to more than 50 feet. One of them, the Geyser, is at least 2,000 feet long, but most of their exposures are much shorter. The vein filling consists generally of calcite, more or less lamellar quartz, and silicified andesite. Most of the gold occurs as fine flakes associated with films and small masses of iron oxide in the vein filling. Workings on most of these prospects consist of shallow shafts and short tunnels, but the Geyser prospect has an inclined shaft 300 feet deep.

SHEEP TANKS DISTRICT[207]

Situation and accessibility: The Sheep Tanks district centers about the Sheep Tanks Mine which is within a mile of the southern margin of the Little Horn Mountains and ¼ mile south of the southeastern corner of T. 1 N., R. 15 W. By way of the Palomas Vicksburg road, the Santa Fe Railway at Vicksburg is 30 miles north, and the Southern Pacific at Hyder is about 35 miles south-southeast of the mine.

---

[206] U. S. Geol. Survey Bull. 620, pp. 160-63. His description is quoted in Ariz. Bureau of Mines Bull. 134.

[207] For a more detailed description, see Ariz. Bureau of Mines Bull. 134, pp. 132-41.

Mining history and production: The first mineral discovery in this vicinity was made by J. G. Wetterhall in 1909. Work in the district was limited to small-scale prospecting until 1926 or 1927 when the Sheep Tanks Mines Company, of Nevada, opened up part of the Resolution vein. Between late 1928 and October, 1929, the Ibex Mines Company, with C. M. d'Autremont as manager, operated this mine. Production for 1929 amounted to 801 tons of siliceous gold ore of smelting grade which contained 1,303.27 ounces of gold, 12,525 ounces of silver, and a little copper, worth in all about $33,514.[208] This ore was hauled by truck to Vicksburg and Hyder and was shipped by rail to the Hayden smelter. For 1929, the property ranked fourteenth among the gold producers of Arizona.

In 1931, the Anozira Mining Company, which later became the Sheep Tanks Consolidated Mining Company, obtained the Resolution ground together with many additional claims and, during part of 1932, carried on extensive prospecting and development work. Water for all purposes was hauled from wells several miles distant. In 1933, three carloads of ore were shipped to El Paso. During the year, a 100-ton cyanide mill was built at the mine. Meanwhile, water was sought by means of shallow wells near camp and a 1,000-foot well in the plain about 3½ miles south of the mine. This deep well, which was started in loosely consolidated sediments, entered volcanic rocks at a depth of a few hundred feet but found sufficient water for the mill. In February, 1934, the plant began treating nearly 100 tons of ore per day. About thirty-five men are employed on the property.

During the winter of 1931-1932, some small pockets of rich gold ore were discovered in the vicinity of the Davis prospect, about 5 miles east of the Sheep Tanks mine. After this discovery, many people located claims in the area and carried on shallow prospecting.

Topography and geology: Here, the basalt mesas of the Little Horn Mountains are interrupted by a northwestward-trending belt, up to about 2 miles wide, of rough, narrow ridges and peaks that rise to a maximum of about 2,800 feet above sea level or 1,200 feet above the adjacent plains. These ridges consist of Tertiary rhyolite, dacite, breccia, agglomerate, and diorite-porphyry. The flows have been affected by complex faulting and fracturing, in part later than the veins.

Resolution vein: The Resolution vein, which comprises the principal ore body of the Sheep Tanks mine, occurs within a fault zone in brecciated andesite. This vein dips northward at a low angle and has been stepped down by later faults. As explored, it is approximately 800 feet long from north to south by

[208] U. S. Bureau of Mines, Mineral resources of the United States, 1929, Part I, p. 827.

700 feet broad from east to west, and ranges from a few inches to about 40 feet in thickness. Its filling consists of irregular masses and streaks of limonite, pyrolusite, quartz of two periods of deposition, and calcite together with more or less silicified diorite-porphyry. In places, irregular, vein-like masses of crystalline barite cut the earlier quartz. All of the Resolution vein contains gold and silver, but the best ore shoot is near the top of its western portion. As exposed by workings, this ore shoot consists mainly of brown and yellow limonitic material, from 2 to 5 or more feet thick, together with vein-like masses of pyrolusite and brecciated fragments, up to 2 feet in diameter, of greenish-yellow quartz. It also contains irregular masses of later, dense, grayish-white quartz and crystalline calcite within vugs and fissures. Locally, small veinlets and blebs of chrysocolla and a green copper-lead mineral are present.

Examined microscopically in thin section, the greenish-yellow quartz is seen to consist of a mosaic of microcrystalline quartz, cut by coarser-grained, comb-like veinlets of quartz and adularia. In places, fragments of silicified rhyolite, mottled by fine specks of hematite, are visible. The greenish-yellow quartz contains small fissures and vugs which are lined with dense to finely crystalline grayish-white quartz and filled with limonite, pyrolusite, and calcite. Some of the fissures and vugs carry chrysocolla and a green copper-lead mineral.

Part of the gold of the Resolution vein occurs within the massive limonitic material, and some is probably contained in the solid portions of the greenish-yellow quartz, but most of the visible gold forms small, thin flakes in iron-stained fractures and vugs within the later, dense, grayish-white quartz.

Wall-rock alteration along the Resolution vein consists of intense silicification of the breccia fragments. Some tens of feet away from the vein, the diorite-porphyry has been extensively chloritized and sericitized, but less silicified. This sericite is very fine grained.

Workings on the Resolution vein include several hundred feet of adit tunnels at various levels, together with open stopes and raises.

Other veins: From the 1,450-foot tunnel that penetrates the ridge beneath the Resolution vein, a 160-foot raise has explored a steeply dipping mineralized fault zone in breccia. This zone, which is from 4 to 6 feet wide, has been largely replaced by pyrolusite, limonite, and calcite. In places, lenticular and rounded masses of greenish-yellow quartz up to a few inches in diameter are present. Microscopic examination of a thin section of this quartz shows it to be similar to the greenish-yellow quartz of the Resolution vein (described on page 144) except that much of it is coarser grained and free of adularia. According to E. Mills, manager of the property, the vein, as exposed in this raise, contains good gold ore in its upper portion and low-grade material in its lower 82 feet above the drift.

On the Smyrna claim, 2,500 feet south of camp, a gold-bearing vein occupies a brecciated zone in diorite-porphyry. This zone strikes northeastward, dips from 30° to 40° SE., and has a maximum exposed thickness of 4 feet. The hanging wall of this vein is marked by a thin streak of coarse-grained gouge. As exposed by a few shallow pits, the vein material consists largely of brecciated porphyry, intermingled with limonite, pyrolusite, and angular to slightly rounded fragments, up to several inches in diameter, of fine-grained, greenish-yellow and white quartz, cemented by calcite, manganiferous calcite, and limonite. Some of the quartz is intermingled with fragments of silicified rhyolite that are cemented by silica. According to Mr. Mills, this vein, as exposed, assays a few dollars in gold per ton.

On the Black Eagle claim, 1,600 feet north of camp, a manganiferous vein occurs along a fault in the breccia. This vein, which is somewhat curved, strikes N. 65° W., dips 45° NE., and is from a few inches to about 2 feet wide. Near the surface, it consists mainly of fine, silicified breccia, cemented by abundant pyrolusite. Fractures in the breccia for 10 or more feet on both sides of the vein, are marked by limonite and manganese stain. As shown by a tunnel, certain cross-fractures are similarly mineralized. At a distance of about 175 feet in from the mouth of the tunnel, the vein is only a few inches wide and consists mainly of manganiferous calcite altering to pyrolusite. No quartz was seen in this vein, but, according to Mr. Mills, the wall rock in places carries a little gold.

The Davis prospect is 5 miles east of the Sheep Tanks mine and a short distance north of the road to Palomas. Here, a wide pediment, mantled by surface gravels, shows a few weathered rock outcrops of andesitic composition. During the winter of 1931-1932, prospecting of these outcrops revealed a brecciated zone that contained a few narrow, lenticular, quartz-carbonate stringers, carrying some gold, and small, limonite-filled pockets that were fairly rich in gold. This quartz, which is of dense texture and pinkish-gray color, contains vugs and cavities lined with white calcite and cut by veinlets of dark, ferruginous calcite. A shallow, inclined shaft, sunk upon this brecciated zone in April, 1932, by the United Verde Extension Mining Company, revealed a few small pockets of rich ore very near the surface. According to George B. Church, the first 20 feet of depth assayed several dollars in gold per ton, but the lower portion of the shaft ran less than $2 per ton. At the bottom of the shaft, a short crosscut and an inclined winze on the zone failed to find ore. These workings indicate that the brecciated zone strikes about N. 55° W. and dips from 30° to 50° N. Except for a few thin lenticular, quartz-carbonate stringers at irregular intervals, this zone is marked only by a thin, wavy, gouge streak and local iron stain. Its brecciated andesite walls have been extensively chloritized and carbonatized and show local sericitization.

About 4 miles east of the Sheep Tanks mine and about ⅛ mile south of the road to Palomas are some shallow workings upon claims held by J. V. Allison. Here, a mass of locally shattered and brecciated diorite-porphyry forms low ridges of eastward trend. A lenticular area of silicified breccia contains prominent, nearly vertical fractures that strike eastward and are marked by considerable amounts of limonitic stain. An old shaft and a few shallow cuts have prospected this brecciated zone. According to Mr. Church, certain portions of the iron-stained, silicified breccia carry a little gold.

Origin of the veins: The veins of this region were originally deposited by hydrothermal solutions which rose along brecciated fault zones. Where these solutions reached the breccia on the Resolution claim, they encountered a flat-dipping fault zone with a relatively impervious cover that caused them to spread outward along the zones of greater permeability. They deposited chiefly manganiferous calcite, gold-bearing quartz of two generations, certain iron and copper minerals, and minor galena and barite. The iron and copper minerals were probably sulphides and may have been auriferous. More or less brecciation of the veins occurred before this deposition was completed. The mineralogy, texture, and wall-rock alteration of these veins clearly indicate that they belong to the epithermal type of deposits.

Subsequent uplift and erosion removed parts of the veins and oxidized the portions now exposed.

Economic possibilitis of Sheep Tanks district: Only a rather limited amount of developed shipping ore remains in the Sheep Tanks region. The district contains considerable amounts of lower-grade material, part of which may be worked at a profit after a cheap water supply adequate for milling needs, has been developed.

In further search for possible undiscovered veins, brecciated zones should receive special attention. Managanese-stained outcrops that contain greenish-yellow quartz with adularia seem to be the most promising but, due to the prevalence of desert varnish in this area, are difficult to recognize from any distance.

Veins of the type that occurs in the Sheep Tanks district are generally not of economic grade below depths of several hundred feet.

TANK MOUNTAINS

Some minor gold-bearing quartz-carbonate veins occur within the granites and schists of the Tank Mountains.

One of the veins carries iron and copper sulphides and all of them contain more or less iron oxide. In general, they are narrow and lenticular with walls that typically show pronounced sericitization, silicification and carbonatization. Such mineralogy and wall-rock alteration suggests deposition in the mesothermal zone.

The Engesser, Blodgett, Golden Harp, Puzzles and Regal prospects of this range are described in Arizona Bureau of Mines Bulletin 134, pp. 124-27.

GILA BEND MOUNTAINS

The Gila Bend Mountains of Yuma County contain several gold-bearing quartz lodes, some of which have yielded a little gold. These lodes generally consist of lenticular or irregular bodies of quartz together with more or less iron oxide. The Bill Taft, Belle Mackeever, and Camp Creek properties are described in Arizona Bureau of Mines Bulletin 134, pp. 145-47.

TRIGO MOUNTAINS

The small farming settlement of Cibola, along the flood-plain of the Colorado River in T. 1 S., is nearly 100 miles by road from Yuma. Ripley, California, its nearest railroad point, is reached by 14 miles of road that connects with Taylor's ferry, on the river.

On the rugged, northwestern slope of the Trigo Mountains, about 6 miles east-southeast of Cibola, are a few square miles of moderately fissile schists, intruded by granite, granite-porphyry, and diorite-porphyry.

A few narrow, branching quartz veins, which occupy fault zones within the schists, have yielded a small amount of gold. The Hardt, Boardway, and Blair properties are described in Arizona Bureau of Mines Bull. 134, pp. 72-73.

CASTLE DOME MOUNTAINS

BIG EYE MINE

The Big Eye Mine, held by A. K. Ketcherside, of Yuma, is on the eastern slope of the Castle Dome Mountains. It is accessible by some 10 miles of road that leads northeastward across the range from Thumb Butte settlement.

As recorded by the U. S. Geological Survey Mineral Resources, this mine from 1912 to 1917 produced $33,185 in gold.

In this region, deep, steep-sided, eastward-trending canyon systems have carved very rugged topography. The mine is on the northeastern slope of a spur that consists mainly of steeply dipping brownish-pink andesitic flows, intruded by dikes of rhyolite-porphyry and apparently faulted against brownish Cretaceous quartzites. These quartzites which form the crest of the ridge west of the mine, strike southwestward, dip steeply southeastward, and are extensively shattered. About ¼ mile south of the mine is a small outcrop of syenite that apparently underlies the Cretaceous rocks.

The vein occurs in a brecciated zone that, at its northern end, strikes south-southwestward and dips steeply westward but, within 150 feet farther south, curves to S. 25° W. and dips steeply westward. This zone is several feet wide. As revealed by surface stopes, the vein is treaceable for about 400 feet. The stopes indicate that the ore shoot was narrow, lenticular, and less than 30 feet in vertical extent. The vein material consists of brecciated, dense, yellowish quartz,, intermingled with veinlets of abundant white crystalline calcite. Microscopic examination of a sample of the vein wall rock shows it to consist of phenocrysts of kaolinized plagioclase in a chloritized groundmass. Many calcite veinlets, bordered by combs of quartz crystals, traverse the rock.

Workings here include, besides the stopes already mentioned, some 800 feet of tunnel that explored the brecciated zone below the stopes.

An old mill, equipped with a crusher, five stamps, and cyanide tanks, is situated about ¼ mile southeast of the mine.

LAS FLORES DISTRICT[209]

Las Flores district, in the southeastern margin of the Laguna Mountains, is accessible by ¾ mile of road that branches northwestward from the Yuma-Quartzsite highway at a point about 3¼ miles from McPhaul Bridge. By this road, the district is approximately 4¾ miles from the railway.

Mineralization here was probably discovered prior to 1865. Raymond[210] stated that, in 1870, "At Las Flores, a small 5-stamp mill has been at work for a part of the year, crushing gold quartz from some small veins in the vicinity. The enterprise seems to be a success . . . " Mexican and Indian placer miners founded the town of Las Flores during the sixties. At present, nothing of this settlement remains except a few ruined adobe buildings, near the highway.

Here, the Laguna Mountains rise to 1081 feet above sea level, or more than 900 feet above the Gila River, and are sharply dissected by short eastward-trending canyon systems. Black, striped, well laminated schists, similar to those in the northern portion of Gila Mountains, make up the prevailing rocks. They generally strike N. 35° E. and dip 20° NW., but some notable local variations, due to faulting, occur. In places, dikes and irregular masses of gray albitic granite intrude the schist and are in turn cut by a few narrow dikes of pegmatite and of fine-grained, white aplitic granite.

The veins occupy zones of shearing and brecciation within

---

[209] For a more detailed description, see Arizona Bureau of Mines Bull. 134, pp. 211-16.

[210] Raymond, R. W., Statistics of mines and mining in the states and territories west of the Rocky Mountains (1870), p. 272. Washington, 1872

schist. The longest is traceable for more than 300 feet, but most of them are of irregular or lenticular form. Some of the more lenticular veins are highly shattered. The quartz is of coarse texture and white to grayish color. In places, its massive structure is interrupted by groups of small cavities. Abundant iron oxide and locally abundant carbonate occur in the cavities and fractures. Sericite and yellow iron oxide, together with minor amounts of manganese dioxide and gypsum, are intermingled with the more shattered quartz. Certain veins contain ragged grains of gold in the quartz and also associated with iron oxide within fractures or cavities. No sulphides occur down to the shallow depth exposed, but their original presence and subsequent oxidation are indicated by the physical and mineralogic character of the veins.

The principal alteration of the vein walls consists of sericitazation along with less marked silicification and carbonatization. These veins clearly belong to the mesothermal type.

Traeger or Agate mine: The old Traeger or Agate mine is on a quartz vein that outcrops a few feet west of a parallel dike of the aplitic granite. The quartz vein, which is 2 feet or more wide, dips westward and extends S. 15° W. for about 300 feet, beyond which it changes in strike and narrows in width. It has been stoped from the surface for a length of about 75 feet and a maximum depth of 25 feet. An inclined shaft extends for some distance down the dip of the vein below this stope. These workings are reported to have been made fifty or more years ago, but no information regarding the grade of the ore and the production is available.

Golden Queen claim: Northeastward, the dike, silicified zone, and quartz veins of the old Traeger mine are interrupted by detrital gravels, but a similar zone, striking also S. 15° W., outcrops about ¼ mile farther north, on the Golden Queen claim. Here, several lenticular quartz veins occupy divergent fault zones. In a low saddle, the principal vein strikes eastward, dips irregularly southward, and is from a few inches to 1½ feet thick. Locally, ragged grains of gold, up to 1/20 inch in diameter, occur in the vein quartz as well as within the iron-stained fractures and the iron oxide cavity-fillings.

A few shallow, old stopes indicate that ore was mined from this vein during the early days. During 1931, a small quantity of rich ore was mined from the outcrop, and a 22-foot shaft was sunk through the principal vein. Early in 1932, an 80-foot shaft was sunk 50 feet north of the vein.

Pandino claim: The Pandino claim, held in 1931 by C. Baker and A. McIntyre, is a few hundred feet northwest of the old Traeger mine. Its workings then consisted of some recent, shallow cuts and an old shaft, probably about 190 feet deep. This shaft, which inclines 30° W. near the surface, is sunk on a granitic dike that is about 3 feet thick and has a thin, lenticular vein of quartz along its footwall. In places, this quartz contains a little free gold.

In June, 1933, several tons of sorted ore were shipped from this claim.

WELLTON HILLS OR LA POSA DISTRICT

The Wellton Hills are a disconnected group of small mountains about 6 miles south of Wellton. They are made up of gneiss and minor amounts of schist, cut by scattered dikes of granite-porphyry and pegmatite. This district contains many low grade, gold-bearing quartz veins within brecciated fault zones which predominately strike between N. 35° W. and W. and dip from 10° to 85° N. or NE. The quartz of these veins is coarsely crystalline, weakly banded, and locally vuggy. It characteristically shows abundant iron oxide together with minor chrysocolla and malachite in irregular bunches or fracture-linings. Brown, crystalline, ferruginous calcite occurs in a like fashion, but more commonly fills the vugs. No sulphides were observed, but the texture and mineralogic composition of the veins indicates that iron and copper sulphides were originally present and have been destroyed by oxidation. The dominant wall-rock alteration accompanying these veins consists of intense sericitization with less marked silicification and carbonatization. Such alteration along veins of this type indicates deposition in the higher temperature portion of the mesothermal zone.

Certain veins of the Wellton Hills are spectacularly marked with copper stain which extends for a short distance into the wall rock. In places, this copper-stained gneiss carries visible specks of free gold. These features, combined with easy accessibility, have, during the last half century, prompted rather active prospecting throughout the area, but very little commercial ore has thus far been developed. Most of the vein exposures and workings are on slopes or saddles rather than on the pediments. The U. S. Geological Survey Mineral Resources for 1909 record from La Posa district a 23-ton test shipment that contained $1,364 of gold, 23 ounces of silver, and 133 pounds of copper, in all worth $1,393.

The Double Eagle, Poorman, Draghi, Donaldson, Wanamaker, McMahan, Welltonia, Northern, and Shirley May prospects, of this area, are described in Arizona Bureau of Mines Bull. 134, pp. 173-76.

LA FORTUNA DISTRICT[211]

La Fortuna district, in the Gila Mountains of southwestern

---

[211] For a more detailed description, see Wilson, Eldred D., Geology and mineral deposits of Southern Yuma County, Arizona: University of Arizona, Arizona Bureau of Mines Bulletin 134, pp. 189-99, 1933.

Yuma County, contains many gold-bearing quartz veins of which only La Fortuna lode has been notably productive.

FORTUNA MINE

Situation and accessibility: The Fortuna mine is at the western base of the Gila Mountains, about 14½ miles southeast of their northern end. It is accessible from the Southern Pacific Railway by some 15 miles of unimproved road that crosses the Yuma-Gila Bend highway at a point 16½ miles from Yuma.

History and production:[212] La Fortuna lode was not discovered until the early nineties, although, nearly fifty years earlier, hundreds of prospectors had traveled along the Camino del Diablo within a short distance of its outcrop. In 1896, Charles D. Lane bought the property for $150,000 and organized La Fortuna Gold Mining and Milling Company[213] which built a 20-stamp mill and operated actively until the close of 1904. The mine and mill employed eighty to one hundred men.

The first four months' run netted $284,600 from ore taken out within 150 feet of the surface. Full production was started in 1898, and, during the following year, a 100-ton cyanide plant was built to treat the accumulated tailings which contained about $5 per ton. In 1900, the vein was lost at a fault on the 800-foot level, and only a small segment was found by further exploration. The total production of the mine from September, 1896, to December,. 1904, was $2,587,987 in bullion sent to the Selby smelter (See table, page 154).

Further exploration was carried on during 1913 and 1914 by the Fortuna Mines Corporation and, from 1924 to 1926, by the Elan Mining Company. These two concerns produced about $25,000 worth of gold. Since that year, the mine has remained inactive.

Topography and geology: The Fortuna topographic sheet, issued in 1929 by the U. S. Geological Survey on a scale of approximately one mile to the inch, includes the Fortuna region. The Fortuna mine is on a narrow dissected pediment between foothill ridges. East of the mine, the range rises very steeply, with sharp, rugged spurs between V-shaped canyons.

In this vicinity, the prevailing rocks are pre-Cambrian or later schist and gneiss, intruded by large masses of granite and amphibolite and numerous dikes of pegmatite and aplite. (See Figure 8.) The schist and part of the gneiss are of sedimentary origin.

As shown by Figure 8, the schist and gneiss strike west or southwest, which is transverse to the trend of the range, and

---

[212] Data compiled by J. B. Tenney.
[213] Eng. & Min. Jour., vol. 93, p. 372, 1912.

dip steeply southward. They have been cut by a network of faults which in places are clearly visible but generally are obscured by talus. As revealed by test pits, the straight gulches follow faults. Due, however, to the rather obscure stratigraphy of the schist, the faulting of this region can be analyzed only after much detailed geologic work.

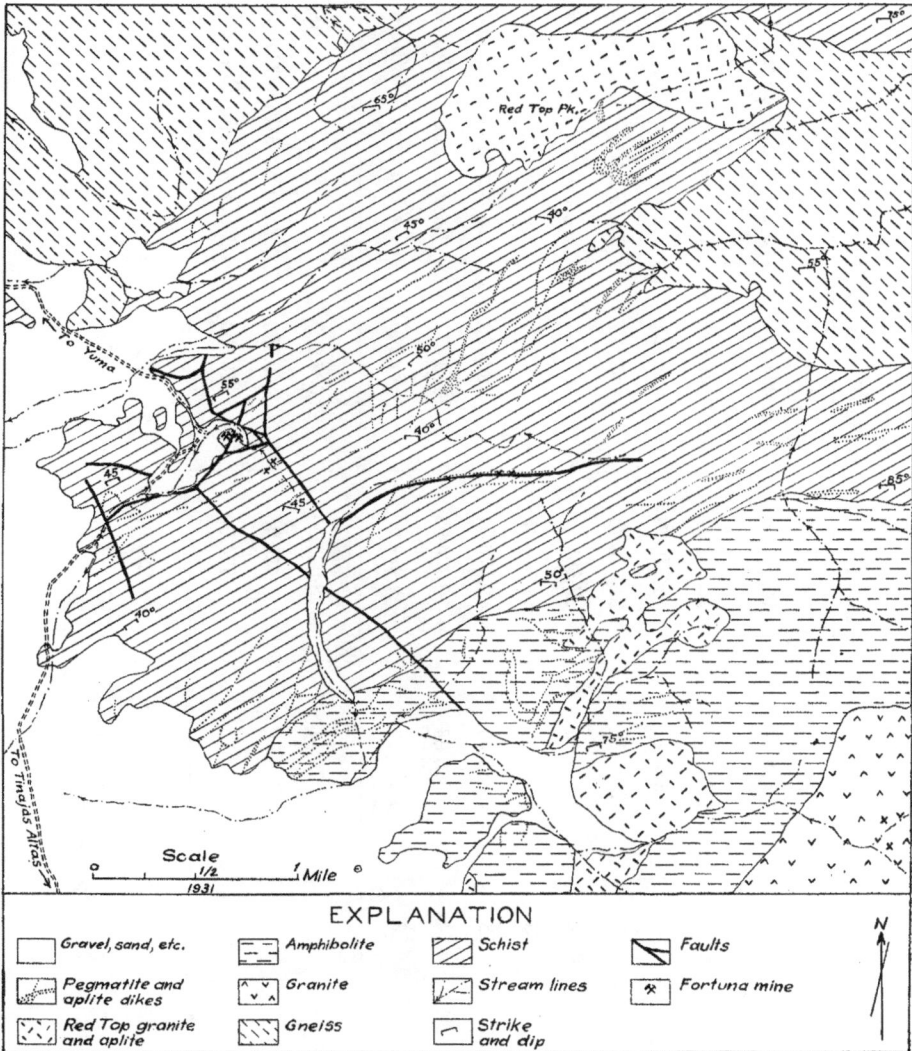

Figure 8.—Geologic map of Fortuna region, Yuma County, 1931.

PRODUCTION, LA FORTUNA MINE*

Date	Price silver	Ounces silver	Ounces Gold	Total value	Remarks
1896	$0.68	1,208.41	14,872.08	$ 308,224	June to December, La Fortuna Mining & Milling Co.
1897	0.60	1,283.09	15,789.92	361,522	La Fortuna Mining & Milling Co.
1898	0.59	1,563.48	15,976.13	331,121	La Fortuna Mining & Milling Co.
1899	0.60	1,595.12	21,078.75	440,770	La Fortuna Mining & Milling Co.
1900	0.62	1,447.40	22,596.58	467,960	La Fortuna Mining & Milling Co.
1901	0.60	783.35	11,994.46	248,411	La Fortuna Mining & Milling Co.
1902	0.53	662.09	9,576.94	198,319	La Fortuna Mining & Milling Co.
1903	0.54	603.88	6,730.95	139,820	La Fortuna Mining & Milling Co.
1904	0.58	1,032.59	4,414.69	91,840	La Fortuna Mining & Milling Co.
Total 1896-1904		10,179.41	123,030.50	$2,587,987	
1913-1914			435.00	9,000	Fortuna Mines Corporation (est.)
1926			774.00	16,000	Elan Mining Company (est.)
Total 1896-1926			124,239.00	$2,612,987	

*Data compiled by J. B. Tenney.

Vein: La Fortuna vein is a branching chimney-like mass that strikes S. 40° W. and dips 70° SE. It outcrops in two branches of which one is 20 feet long by 12 feet wide and the other 30 feet long by 5 feet wide. These branches are reported to have joined at a depth of 500 feet, forming an ore body approximately 100 feet long by 1½ to 12 feet wide. According to F. J. Martin,[214] manager of the property during most of its activity, the ore down to the 200-foot level averaged more than $30 per ton, and everything milled averaged between $15 to $16 per ton. The gold was about .890 fine. The ore body was lost, as has been stated, at a fault near the 800-foot level, and only a small segment, between the 900- and 1,100-foot levels, was found by further exploration.

As seen in specimens, the gold-bearing quartz is coarse grained, vitreous, pale straw colored, and locally stained with malachite. Examined microscopically in thin section, it is seen to consist of irregular interlocking grains, up to 0.15 inch long by 0.07 inch wide.

Microscopic examination of a polished section of the high-grade quartz shows minute bodies and small irregular to interlacing veinlets of hematite, locally altered to limonite. The gold appears as round grains within converging hair-like cracks and also as thin irregular veinlets within the limonite.

The wall-rock alteration accompanying La Fortuna vein consists mainly of carbonatization and silification. Such alteration, together with the structure, texture, and mineralogy of the vein, point to deposition in the lower portion of the mesothermal zone.

Workings: The Fortuna mine workings include two inclined shafts which connect with several hundred feet of drifts, stopes, etc. The older shaft, which is on the ridge above the mill and 250 feet southwest of the vein outcrop, inclines 60° in a N. 34° E. direction and is 350 feet deep. The lower shaft, which is 100 feet southeast of the outcrop, inclines 58° in a S. 54° E. direction and is approximately 1,000 feet deep. In 1929, these shafts were caved at the surface, and the workings were reported to be partly filled with water.

Mill: The Fortuna 20-stamp mill[215] and 100-ton cyanide plant have been practically dismantled.

MINOR GOLD DEPOSITS

Many quartz veins cut the schist in the vicinity of the Fortuna mine. In general, this quartz is coarse grained, white, and locally copper stained, but lacks the straw color characteristic of the Fortuna vein. Most of these vein outcrops have been prospected by tunnels and shallow shafts, some of which now serve as cis-

---

[214] Written communication.
[215] Described in Ariz. Bureau of Mines Bulletin 134, pp. 196-99.

terns for rain water. A little gold has been found in some of the veins, but no production is reported from any of them.

On the crest of the Gila Mountains, 3 miles north of the Fortuna mine, a little prospecting has been done on quartz veins in gneiss, but operations have been greatly hampered by the ruggedness of this part of the range. These veins generally contain more pulverent, red to black, iron oxide than quartz and have irregular widths of less than one foot. Some of them outcrop over lengths of several hundred feet, and one is traceable for about ½ mile. The quartz is coarse and even grained but broken by many fractures that are filmed with iron oxide. In places, thin, fine flakes of gold are abundantly scattered over the fracture surfaces, and sparse rounder particles are within the more solid quartz. Small grains of pyrite are present in the quartz. A little sericite occurs in the immediately adjacent wall rock.

Certain quartz veins in the northern portion of the Gila Mountains have been found to contain small amounts of gold, but little or no production has been made from them.

ECONOMIC POSSIBILITIES OF LA FORTUNA DISTRICT

The faulted segment of the Fortuna vein may eventually be found, particularly if future exploration for it is guided by thoroughly accurate, detailed stratigraphic and structural studies of the area.

Despite the fact that much search has been made for possible undiscovered gold-quartz veins in this region, further prospecting is warranted. The schist offers the most possibilities from the standpoints of permeable zones and structure, but none of the formations can yet be excluded as barren terrain. If the Fortuna vein is genetically connected with any of the stocks of the Red Top granite shown on Figure 8, the area for a few miles around these stocks is favorable ground. The best possibilities are along the margins of the range, on the pediment, particularly where the outcrops are hidden by gravels or talus.

CHAPTER VI—MARICOPA COUNTY

Maricopa County, as shown by Figure 9, comprises an irregular area about 130 miles long by 105 miles wide. It consists of broad desert plains with scattered mountain ranges that, for the most part, are made up of pre-Cambrian schists and granites and Tertiary volcanic rocks.

This county, which ranks fifth among the gold-producing counties of Arizona, has yielded approximately $7,400,000 of gold, most of which has come from the Vulture mine.[216]

[216] Statistics by J. B. Tenney.

VULTURE DISTRICT

VULTURE MINE

Situation: The Vulture mine is at the southern margin of the Vulture Mountains, about 9 miles west of the Hassayampa River and 14 miles by road southwest from Wickenburg.

History:[217] The story of the discovery of this deposit is given by Browne[218] as follows:

"A German, named Henry Wickenburg, with several companions, while prospecting upon the Hassayampa late in 1863, discovered a butte of quartz . . . After examining it closely they found traces of gold but attached no great value to the ore, and all but Mr. Wickenburg were reluctant to go to even the slight trouble of posting notices to claim the lode." During the next three years, Wickenburg treated rich portions of the outcrop ore in an arrastre at the river. The activities of the Apaches probably handicapped his operations.

Late in 1866, the Vulture Company, of New York, acquired the property, established a camp at the mine, and built a 40-stamp amalgamation and concentration mill near the site of the present town of Wickenburg. All of the machinery was shipped by water from San Francisco to Fort Mohave, a landing on the Colorado River, and hauled overland via Prescott.

This company operated steadily from 1867 until July, 1872, when the apparent pinching of the ore at water level and the $8 to $10 per ton charge for freighting ore from mine to mill discouraged the owners. During this period, approximately $1,850,-000 worth of bullion was obtained from ore that ranged from $25 to $90 in gold per ton. More than 6,000 tons of concentrates and 80,000 tons of tailings that averaged $5 per ton were stored. Mining, milling and hauling costs amounted to $14.93 per ton.[219] About one hundred and twenty-five men were employed at the mine and mill.

In 1873, P. Smith and P. W. Taylor located a claim on the western extension of the lode and built a 5-stamp mill at the Hassayampa River. They operated intermittently for six years and produced about $150,000 worth of bullion.

In 1879, the Arizona Central Mining Company was formed to work the Vulture and the Taylor-Smith claims. An 80-stamp mill was built at the mine and connected with the Hassayampa River by a pipe line. This company operated on a big scale for nine years and treated a large amount of low-grade ore. Exact production figures for this period are lacking, but scattered es-

---

[217] Abstracted from unpublished notes of J. B. Tenney.
[218] Browne, J. Ross, Mineral resources of the states and territories west of the Rocky Mountains, p. 477, Washington, 1868.
[219] Raymond, R. W., Statistics of mines and mining in the states and territories west of the Rocky Mountains, p. 260. Washington, 1872.

KEY TO MINING DISTRICTS SHOWN ON FIGURE 9

MARICOPA COUNTY DISTRICTS

1 Vulture
2 Big Horn
3 Midway (Saddle Mountain)
4 White Picacho

5 Agua Fria
6 Cave Creek
7 Winifred
7-A Salt River

GILA COUNTY DISTRICTS

8 Payson or Green Valley
9 Spring Creek or Young

10 Globe
11 Banner or Dripping Spring

PINAL COUNTY DISTRICTS

12 Goldfields
13 Superior (Pioneer), Mineral
     Hill
14 Saddle Mountain
15 Cottonwood

16 Mammoth (Old Hat)
17 Casa Grande
18 Old Hat
19 Owl Head

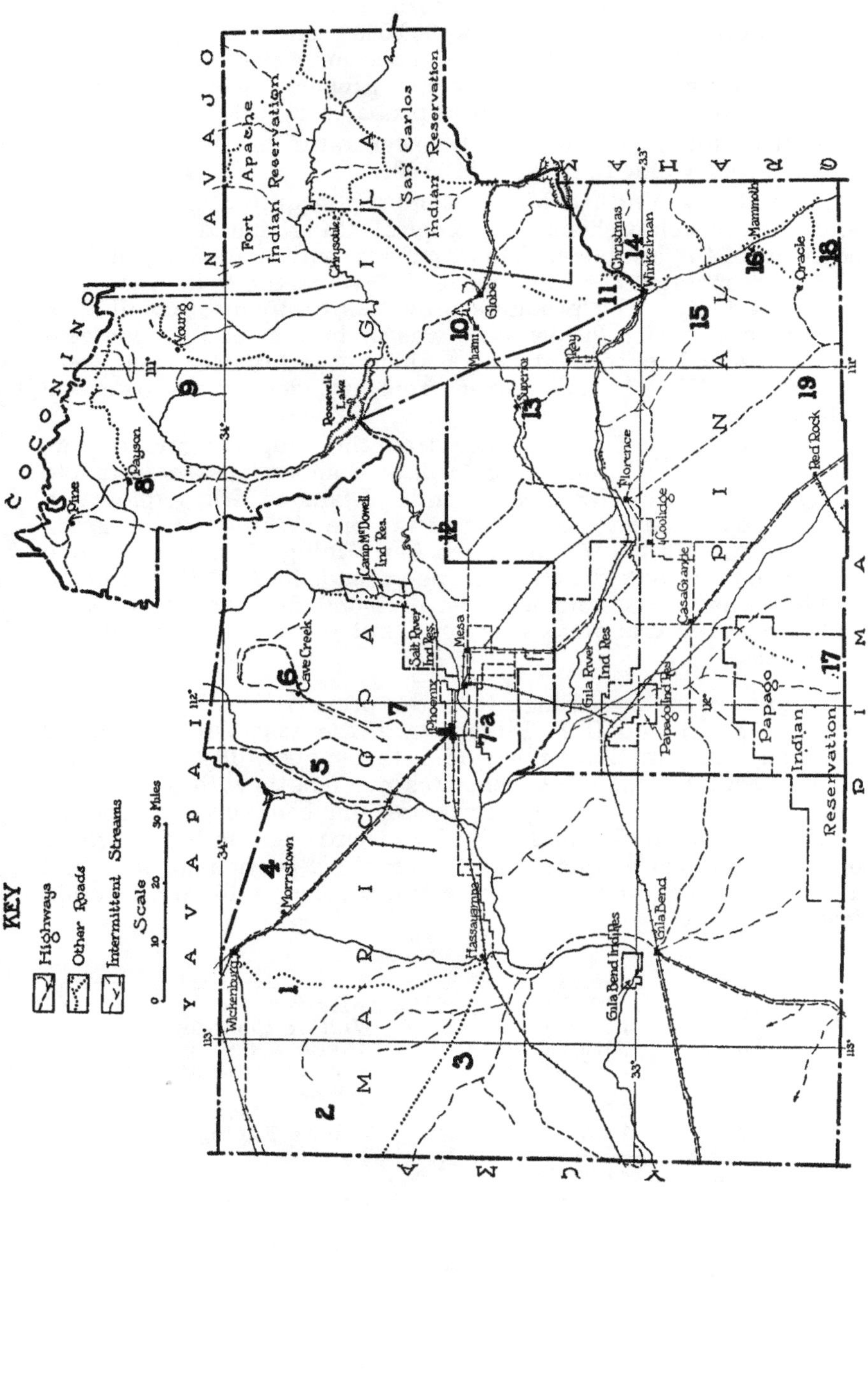

Figure 9.—Map showing location of lode and gold districts in Maricopa, Gila and Pinal counties

timates by the Arizona Daily Star and U. S. Mint reports indicate a probable yield of about $2,000,000. The ore body was lost at a fault on the 300-foot level, and the mine was closed in 1888. During several ensuing years, the property was worked by lessees who made a production of probably $500,000.

In 1883, shipments of the old concentrates and tailings of the original mill yielded probably $500,000.

In 1908, the Vulture Mines Company acquired the property and, after a comprehensive geological study, found the faulted segment of the ore body. This company built a 20-stamp mill in 1910 and operated the mine until 1917 when the vein was again lost at a fault. The production by this company amounted to $1,839,375 of which 70 per cent was in bullion and 30 per cent was contained in concentrates. About two hundred men were employed. Water was pumped from two deep wells near the mine.

In 1927, D. R. Finlayson acquired the property and organized the Vulture Mining and Milling Company. Ore from old pillars was treated in a 5-stamp mill. Diamond drill exploration for the second faulted segment of the ore obtained encouraging results. The United Verde Extension Mining Company became financially interested in the property and, in 1930-1931, sank a 500-foot shaft to supplement the diamond drilling. More than 1,000 feet of lateral work was done, but the results were disappointing.

Recent operations: Since 1931, the property has been worked by A. B. Peach and D. R. Finlayson, of the East Vulture Mining Company. From September, 1931 to October, 1933, they produced about 10 tons of concentrates per month with a 10-stamp mill. When visited in February, 1934, this company was operating a 125-ton amalgamation and concentration mill for which ore was obtained by quarrying the unmined portions of the vein. The old tailings dump was being run through a 100-ton cyanide leaching plant.

PRODUCTION SUMMARY[220]

1866 to 1872	$1,850,000	Vulture Company
1873 to 1878	150,000 (est.)	Taylor & Smith
1873 to 1890	1,000,000 (est.)	Lessees; ore and old concentrates.
1879 to 1888	2,000,000 (est.)	Arizona Central
1908 to 1917	1,839,375	Vulture Mining Company
Total..................	$6,839,375	

[220] Figures compiled by J. B. Tenney.

Topography and geology: The high southeastern portion of the Vulture Mountains is made up of andesitic and rhyolitic lavas which lie upon a basement of schist and granite. In places, granite and rhyolite-porphyry dikes are abundant. The Vulture mine is near the southern or outer margain of a moderately hilly pediment, at an altitude of 2,000 feet. This pediment is floored with quartz-sericite schist, intruded by granite and rhyolite-porphyry. Complex faulting, partly pre-mineral and partly post-mineral, has affected these formations.

Vein and workings: The Vulture vein occurs within a fault zone that, at the surface, strikes slightly north of west and dips 45° N., nearly parallel to the lamination of the schist in the footwall. The hanging wall is partly a granite-porphyry dike, up to 80 feet wide, and partly schist. Near the vein, these rocks contain abundant sericite and some calcite and pseudomorphs of pyrite metacrysts.

Raymond,[221] who visited the mine in 1870 or 1871, says: "The croppings of this remarkable lode rise 80 feet above the level of the mesa . . . Eighty-five feet in width of this is vein matter which lies between well-defined walls. These croppings at the surface show gold everywhere; but there are here four distinct quartz layers which are richer than the remainder and have the following widths: The 'Red' or 'Front' vein, 12 feet; the 'Middle' vein, 6 feet; the 'Blue' vein, 9 feet; and the 'Black' vein, 5 feet; total width, 32 feet. These are not mined, but quarried, all above the level of the mouth of the main shaft being taken down together. Even in the talc (sericite) slate horses, between the pay-quartz, is gold . . . (In the slate) there are also numerous small quartz seams, from an inch to one foot thick, which contain much gold.

"At the 240-foot level the thickness of the vein is 47 feet. The richest ore lies here nearest to the walls."

The typical vein quartz is coarsely crystalline, locally cellular, and grayish white to white. Hutchinson[222] says: "In the oxidized zone the quartz is stained with iron oxide, and some wulfenite in characteristic tabular crystals is found in openings in the quartz . . . Below the zone of oxidation the vein minerals, other than quartz, are pyrite, galena, blende, and chalcopyrite. The proportion of these is indicated by the ratio of concentration, which was about thirty to one, and the assay of the concentrates, which was 12 to 15 per cent of lead, 8 to 12 per cent of zinc, one to 2 per cent of copper, and from $120 to $200 in gold. Metallic gold was found in all parts of the mine. Even in the deeper

[221] Raymond, R. W., Statistics of mines and mining in the states and territories west of the Rocky Mountains, pp. 257-58, Washington, 1872.

[222] Hutchinson, W. S., The Vulture mine: Eng. and Min. Jour., vol. 111, no. 7, pp. 298-302. Feb. 12, 1921.

workings where the ore was not oxidized but was made up of characteristic quartz with associated sulphides, coarse gold was present . . .This gold had a fineness of 760 to 780 . . .The galena was usually rich, so that, when the average mill concentrates assayed $150 per ton, the clean galena concentrate assayed $600.

"The outcrop was 1,000 feet long, but . . . the upper parts of the vein have been quarried in two large open pits. The westerly pit is 300 feet long and the easterly one 500 feet, with low-grade vein matter, which consists mostly of white quartz, remaining between them."

As indicated by areas of stoping shown on maps of the mine workings, the quarry pits were on the outcrops of two steeply eastward-pitching ore shoots of which the western one was mined to the 600-foot level, and eastern to the 1,000-foot level. Westward, the vein extends into granite and splits into several small but locally rich branches. Hutchinson continues:

"Granite of identical character was encountered in the westerly end of the 950 level, in the easterly end of the 1,550 level, and in a diamond drill hole put down from the latter. These points of exposure of granite indicate a probable easterly pitch of the contact."

Besides numerous faults of small displacement, two large faults, the Talmadge and Astor, have cut the vein. Hutchinson states that the Talmadge fault, which cuts the vein above the 450-foot level of the east shaft, dips 80° NE. and has a vertical displacement of 300 feet. The Astor fault, which cuts off the vein below the 950-foot level, is reported to be nearly parallel to the Talmadge fault, but its displacement remains unknown. Cross-sections of these features are given by Hutchinson, in the article already cited, and by A. P. Thompson, in Min. Jour., vol. 14, pp. 9-11, 28-30, 1930.

SUNRISE MINE

The Sunrise mine is in northwestern Maricopa County, about 18 miles west of Wickenburg and 2½ miles south of U. S. Highway 60.

This deposit was located in 1915. In 1927, it was purchased by W. M. Ebner and associates who sank a 330-foot incline and did about 2,000 feet of development work. C. W. Mitchell obtained the property late in 1933 and, from March 1 to May 16, 1934, shipped 600- tons of ore that averaged $24 in gold per ton.[223] About fourteen men were employed. Water for all purposes is hauled from Aguila, 11 miles distant.

The mine is at the southern base of some low hills that are composed of schist intruded by granitic porphyry. The vein strikes S. 20° W., dips about 45° NW., and occurs within a fault zone with granitic porphyry on the hanging wall and schist on

---

[223] Oral communication from Mr. Mitchell.

the foot wall. The vein is a stockwork, from 10 to 20 feet wide, of lenticular quartz veins, from a few inches to a few feet thick, in schist. Its outcrop is largely mantled by detritus.

The main adit or 200-foot level includes about 600 feet of drifts, and the 330-foot level about 150 feet of drifts. Most of the stopes extend above the 200-foot level. At the time of visit, the largest stope was some 45 feet high by 15 to 20 feet long by 4 to 5 feet wide.

The ore shoots appear to occur where the vein flattens and is intersected by transverse fractures. The ore consists of coarse, locally honeycombed to platy, brecciated white quartz with abundant limonite and hematite. In isolated places, a little pyrite is present. Most of the gold occurs as mediumly fine to coarse grains and flakes, mainly with pinkish-red hematite and limonite in fractures and cavities. The honeycombed and platy quartz with the hematite and limonite is reported to be of particularly high grade. According to Mr. Mitchell, the ore contains less than 0.25 ounce of silver per ounce of gold.

Wall-rock alteration along this vein consists of sericitization, silicification, and carbonatization.

BIG HORN DISTRICT

EL TIGRE MINE

El Tigre property of twelve claims, in the northwestern Big Horn Mountains mining district, of northwestern Maricopa County, is 15 miles by road south of Aguila.

This deposit was located in 1914 by the Sisson Brothers. According to local people, it was worked mainly between 1918 and 1924. During 1921, some bullion was produced in a 10-stamp mill built near a well, 3½ miles west of the mine. In 1922, ore was run through this mill, and old tailings were treated by cyanidation. According to J. B. Webb, the January, 1923, yield amounted to $14,454 worth of gold.[224] Figures on the total production are not available.

At the mine, fine-grained gneissic granite, intruded by basic dike rocks, floors a hilly pediment. The ore, which occurs within a nearly flat fault zone, consists of massive to coarse-grained shiny quartz with abundant specularite and limonite. The wall rock has been notably altered to sericite.

Most of the production came from drifts and stopes which extend for a few tens of feet into the vein. These workings indicate that the ore body was very lenticular, with a maximum width of about 5 feet. Three inclined shafts, 50, 197, and 200 feet deep, respectively, were sunk below the outcrop. They are reported to have cut two separate veins, but little or no production was made from them.

---

[224] Oral communication.

CAVE CREEK DISTRICT

The Cave Creek district is from 25 to 45 miles by road north of Phoenix. It is in a group of low, moderately rugged mountains composed of schist, gneiss, and granite and overlain on the north by volcanic rocks.

This district contains deposits of copper, gold, silver, lead, tungsten, molybdenum and vanadium, but only the copper and gold deposits have been notably productive. The gold production, which probably amounts to about $250,000, was mostly made prior to 1900. During recent years, the district has yielded only a few thousand dollars' worth of gold.

The largest early-day gold producers of the district were the Phoenix and Maricopa properties, which are a few miles north of Cave Creek post office, on the gentle slope east of the main creek. Their country rock is altered schist, intruded by dikes of granite-porphyry and rhyolite-porphyry. The gold was apparently associated with fine-grained, grayish quartz in silicified brecciated zones of northward strike and steep dip.

The *Phoenix* mine was opened by several thousand feet of workings that are now largely caved but are reported to be distributed over a width of about 300 feet and to extend to a depth of 90 feet. According to A. S. Lewis,[225] it produced about $100,000 worth of bullion with a 20-stamp mill and, in the nineties, was equipped with a 100-stamp mill and a cyanide plant.

The *Maricopa* property, which joins the Phoenix on the south, is reported to have been opened by some 600 feet of workings that extended to a maximum depth of 100 feet. Its present owner, A. S. Lewis, states that, prior to 1900, it produced about 5,000 tons of $15 ore. He states that the ore shoot was from 6 to 8 feet wide and contained, besides gold, about 0.5 per cent of molybdenum and a little vanadium.

The *Mormon Girl* mine, one mile south of Cave Creek, is on a notable old copper deposit that, in places, contained considerable gold. It is reported to have yielded three carloads of ore in 1931. According to A. S. Lewis,[225] the ore from the 300-foot level contained 5 per cent of copper and 0.5 ounce of gold per ton.

The *Copper Top* mine, 2½ miles southwest of Cave Creek, is on a copper-gold-silver-lead deposit that is reported to have produced some rich gold ore during the early days. A. S. Lewis[225] states that it workings include two shafts, 300 and 100 feet deep.

The *Mex Hill* property, on upper Rowe Wash, is reported to have yielded about $3,000 worth of rich gold ore during the early days. Recently, U. Moss has explored this ground with a 350-foot tunnel.

The *Lucky Gus* or old *Ed Howard* property, near the Grapevine road, northeast of Cave Creek, made a small production of

---

[225] Oral communication.

rich gold ore several years ago. In May, 1934, it was being explored by a tunnel below the old 50-foot shaft.

The *A. B. Bell* property, at the head of Blue Wash, 12 miles northeast of Cave Creek, is reported to have yielded about $2,000 worth of gold during the early days. It was operated for a few months during 1933 and produced a few hundred dollars worth of concentrates and amalgam with a 5-ton mill.[226]

The *Rackensack* mine, owned by L. E. Hewins, is in Rackensack Gulch, about 4 miles upstream from the Camp Creek highway bridge. In May, 1934, this mine was being worked through a tunnel by A. Verkroost. During the past three years, it has yielded more than $1,000 worth of ore.[226] This ore was packed for about 2 miles to the Dallas-Ft. Worth mill.

The *Dallas-Ft. Worth* property, now under option to Charles Diehl and Crismon Bros., is in Rackensack Gulch, about 2 miles upstream from the highway bridge. This property is said to have produced a few thousand dollars' worth of gold during the nineties. It is equipped with a small stamp mill.

The *Gold Reef* property, held by Dan Steele, is a few miles northeast of Cave Creek and 38 miles by road from Phoenix. In May, 1934, the Stuart Gold Reef Mines, Inc., was working the property and operating a 5-stamp unit of an old 10-stamp mill. The ore was being obtained from open cuts on a gently dipping vein near the top of the mountain, some 1,500 feet above the mill. Nine men were employed. This vein occurs in schist, east of a large stock of reddish granite. The ore is cellular, milky-white quartz with some black hematite. The gold, which is fine grained, occurs mainly in the cavities.

WINIFRED DISTRICT

JACK WHITE MINE

The Jack White mine is in the northern foothills of the Phoenix Mountains, about 18 miles by road from the railway at Phoenix.

This deposit was located during the eighties. In 1913, J. White and associates organized the Eyrich Gold Mining Company which sank the shaft to a depth of 300 feet and ran some drifts. C. K. Barnes erected a 10-stamp mill on the property in 1928 and produced several thousand dollars' worth of bullion. In 1931, the Hartman Gold Mining and Milling Company sank the shaft to the 500-foot level, did considerable drifting, and shipped several car loads of ore that contained from $12 to $16 worth of gold per ton.[227] A new mill, equipped for flotation and concentration, was built in 1932, but operations were suspended in October, 1933. When visited in May, 1934, the mine was being worked on a small scale by Mr. White.

---

[226] Oral communication from A. S. Lewis.
[227] Oral communication from J. White.

The principal rocks in this vicinity are dark-gray granite with small included masses of schist and a few dikes of acid porphyry.

The vein, which is traceable for a few hundred feet on the surface, strikes southwestward at its northern end, southward at its southern end, and dips about 60° W. It has been opened by a 500-foot inclined shaft and about 3,000 feet of drifts. An ore shoot seen on the 200-foot level is about 35 feet long by a maximum of 2½ feet wide but lenses out abruptly. The ore consists of coarse-textured, locally vuggy, grayish-white quartz with some calcite and irregular bunches of hematite. In a few places, unoxidized masses of finely granular pyrite are present. The gold is finely divided and has formed no placer deposits.

According to Mr. White, one stope between the fourth and fifth levels is about 50 feet long at the top by 100 feet long at the bottom and averages about 2 feet in width. He states that the three ore shoots exposed underground pitch steeply southward.

SALT RIVER DISTRICT

MAX DELTA MINE

The Max Delta mine, held by T. C. McReynolds, Jr., is in the northern portion of the Salt River Mountains, about 9 miles by road south of the railway at Phoenix.

This deposit was located prior to 1900. Several years ago, while being developed by a 500-foot shaft and extensive drifts, it produced a small tonnage of gold ore which contained a little silver. Since late 1933, the mine has been actively operated by the Ace Mining and Development Company. Up to May 10, 1934, this company shipped about 1,600 tons of ore that averaged 0.70 ounce of gold and approximately one ounce of silver per ton.[228] About twenty-four men were employed. The costs of trucking the ore 9 miles to the railway amount to 75 cents per ton.

The principal workings are near the northern base of the range, in schist and gneiss that dip prevailingly northward at low angles but show much local disturbance by faulting. Most of the underground work has been done on three intersecting, faulted veins of which one strikes N. 60° W. and dips 50° NE., another strikes N. 30° W. and dips 60° NE., and the third strikes N. 10° W. and dips 20° to 40° NE. Workings on these veins include about 2,500 feet of tunnels, besides some stopes. The 500-foot inclined shaft already mentioned, which is on the vein of N. 60° W. strike, contained about 100 feet of water in May, 1934.

Most of the ore mined has come from faulted segments of the vein of 20° to 40° dip, particularly from the vicinities of its intersections with other veins. Due to the complex faulting, detailed mapping of the various structural features has been neces-

---

[228] Oral communication from W. M. Snow, of the Ace Mining and Development Company.

sary in prospecting. As indicated by areas of stoping, the largest ore shoot was about 100 feet long by 100 feet high on the dip by an average of 5 feet wide. The ore shoots are lenticular veins of quartz that locally form stockworks up to 8 feet wide. As these stockworks are commonly about half country rock, the ore is hand sorted before shipment. The ore consists of coarse-textured grayish-white quartz with irregular bunches and disseminations of pyrite, locally altered to limonite and hematite. It generally shows no gold when panned and has formed no placers. The wall rocks shows considerable sericitization, but the vein outcrops are very inconspicuous.

CHAPTER VII—PINAL COUNTY

Pinal County, as shown by Figure 9, (page 159), comprises an irregular area about 100 miles long by 65 miles wide. It consists of broad desert plains with scattered mountain ranges of pre-Cambrian schist, granite, and sedimentary rocks, Paleozoic sedimentary beds, and Tertiary volcanic rocks. In places, particularly in the eastern part of the area, stocks of Cretaceous to Tertiary granite and granite-porphyry occur.

Prior to 1932, this county, which ranks sixth among the gold-producing counties of Arizona, yielded approximately $5,474,000 worth of gold of which about $3,120,000 worth came from lode gold mines.[229] Most of this production was made by the Mammoth district, in the southwestern part of the county.

GOLDFIELDS DISTRICT

YOUNG OR MAMMOTH MINE

The Young or Mammoth property, in the Goldfields district, of northern Pinal County, is between the Superstition and Goldfield mountains, 36 miles east of Phoenix via the Apache Trail. This deposit is reported to have been worked during the early nineties by C. Hall and D. Sullivan, but little of its early history and production are known. Blake,[230] in 1898, stated that the mine had been extensively worked to a depth of 100 feet. A considerable production was made from the "Mormon stope," but the amount is unknown. The Young Mines Company, Ltd., headed by G. U. Young, acquired the property in 1910 and spent about fifteen years exploring it with three shafts and thousands of feet of drifts. The main shaft was sunk to a depth of 1,000 feet. Numerous buildings and other surface equipment, including a large steam-driven electric power plant, were erected. During part of this period, a 10-stamp amalgamation mill and a 50-ton cyanide plant were operated intermittently and yielded about

---

[229] Statistics compiled by J. B. Tenney.
[230] Blake, Wm. P., in Rept. of Gov. of Ariz., 1898, p. 260.

$67,000 worth of gold and silver from 7,100 tons of ore.[231]   The
Apache Trail Gold Mining Company, a reorganization of the
Young Mines Company, Ltd., made a small production during
1929 and 1930.   When visited in February, 1934, the 1,000-foot
level was reported to be largely caved and all of the workings
below the 200-foot level were under water.   The Goldfield Mining
and Leasing Company, lessees, were preparing to unwater the
mine.
   This area is a pediment floored by coarse-textured granite,
indurated conglomerate, and granite breccia.   The principal
workings are reported to have been in the vicinity of two northward-trending, steeply westward dipping faults which outcrop
some 300 feet apart.   The most productive ore body, that of the
"Mormon stope," which was mined out prior to 1898, occurred
north of the main shaft, at the intersection of a cross-fault with
a sheeted zone.   The caved portion of this stope is about 100 feet
long by 25 feet wide.   Here, the granite outcrop is heavily stained
with brownish limonite and contains a few irregular stringers of
coarse-grained white quartz.   According to George B. Church,[232]
of the Goldfield Mining and Leasing Company, all of the material
mined was oxidized.   The gold-bearing iron oxide probably was
derived from pyrite which accompanied quartz.

PIONEER DISTRICT

   The Pioneer district, in northeastern Pinal County, in the vicinity of Superior, has been noted mainly for the Magma copper
mine and the old Silver King silver mine.   Prior to 1932, this
district yielded approximately $2,056,000 worth of gold, most of
which was a by-product of copper mining.   During the last two
years, lessees have produced more than $270,000 worth of gold
ore.
   Superior is at an altitude of 3,000 feet, at the western base of
a steep, rugged slope known in part as Apache Leap.   This slope
is extensively dissected by westward-trending gulches of which
Queen Creek Canyon, southeast of Superior, is the largest.
   The following succession of rocks is present in this vicinity:[233]
Pre-Cambrian Pinal schist; pre-Cambrian or Cambrian conglomerate, shale, quartzite, and limestone, intruded by extensive sills
of diabase; Cambrian quartzite; Devonian, Mississippian, and
Pennsylvanian limestones intruded by dikes and sills of quartz
monzonite-porphyry; and Tertiary dacite flows.

---

[231] Figures compiled by J. B. Tenney.
[232] Oral communication.
[233] Short, M. N., and Ettlinger, I. A., Ore deposition and enrichment at the
    Magma mine, Superior, Arizona: Am. Inst. Min. Eng., Trans., vol. 74,
    pp.174-189, 1926; Ransome, F. L., Copper deposits near Superior, Ariz.:
    U. S. Geol. Survey Bull. 540, pp. 139-46, 1914; Darton, N. H., A résumé
    of Arizona geology: Univ. of Ariz., Ariz. Bureau of Mines Bull. 119,
    pp. 269-72, 1925.

Before the outpouring of the dacite and immediately after the intrusion of the Central Arizona batholith, as interpreted by Short and Ettlinger, the region was broken by eastward-trending faults and the hypogene copper deposits were formed. After long erosion, the formations were uplifted and tilted approximately 35° eastward, and the region was buried by 1,000 feet or more of dacite flows. Later uplift, accompanied by northward-trending faults of great magnitude, gave rise to the present mountain ranges.

LAKE SUPERIOR AND ARIZONA MINE

The Lake Superior and Arizona mine is at the eastern edge of Superior, between Queen Creek and the Magma mine.

During the early days, the Gold Eagle mine produced some gold and silver-bearing copper ore from this ground. In 1902, the property was obtained by Michigan capitalists who organized the Lake Superior and Arizona Mining Company. This company drove a few thousand feet of tunnels and sank a 1,400-foot incline. According to Weed's Mines Handbook, a production of 99,120 pounds of copper, 1,040 ounces of silver, and 188 ounces of gold was made in 1907. During the World War, lessees mined some copper ore from the property. Since 1920, the ground has been owned by the Magma Copper Company.

In 1932, T. D. Herron and C. Laster leased the mine and opened large bodies of gold ore. To the end of 1933, these lessees shipped nearly 7,000 tons of ore to the Magma smelter. During the first five months of 1934, their production of gold and silver amounted to more than $100,000. In May, 1934, fifty men were employed.

The rocks in the vicinity of the Lake Superior and Arizona mine are Cambrian quartzite and Devonian limestone which strike northward and dip about 30° E. The vein occurs within the zone of a strike fault that has brecciated the quartzite-limestone contact and the lower beds of the limestone. As stated by Ransome,[234] this brecciation is associated in surface exposures with limonite, manganese oxide, quartz, and hematite, and in places with malachite and chrysocolla.

More than twenty-five years ago, this zone was opened by a shaft, some 1,400 feet deep, that inclines 26° E. and connects with eight levels of drifts. Most of the drifting is on the second or Carlton tunnel level which extends southward for some 2,000 feet and opens into Queen Creek Canyon. East of its portal is an old vertical shaft, the Vivian, that taps the vein at a depth of 140 feet. These old workings, which were mainly in the foot-wall portion of the vein, exposed only a few small bodies of oxidized copper ore, and material that contained generally less than 0.2 ounces of gold per ton.

The present lessees, by cross-cutting along transverse fissures for a few feet towards the hanging wall, discovered five shoots

---

[234] Work cited, p. 155.

of gold ore within a horizontal distance of 3,000 feet. These ore bodies average 4 feet wide by 15 feet long, and the most persistent one extends, with interruptions, to the bottom level of the mine. The ore consists mainly of hematite, limonite, and fine-grained grayish to greenish-yellow quartz of epithermal aspect. As a rule, the gold is spongy to fine grained and occurs erratically distributed. According to Mr. Herron,[235] the ore mined contained generally an ounce or more of gold and an ounce of silver per ton.

QUEEN CREEK MINE

The Queen Creek mine is a short distance south of Queen Creek, on the southward extension of the Lake Superior and Arizona vein. Its workings include an 817-foot inclined shaft with several hundred feet of drifting on seven levels south from the shaft. Since early 1933, it has been operated by G. J. Poole and E. Cockerhan, lessees, who, prior to June, 1934, shipped about seven cars of ore to the Magma smelter. The last three cars are reported to have contained about one ounce of gold and 2 ounces of silver per ton.[236]

BELMONT PROPERTY (SMITH LEASE)

The Chas. H. Smith lease, on the Monte Carlo claim of the Belmont Copper Mining Company property, is some 1,500 feet northeast of the Belmont mine and 4 miles southeast of Superior. From December, 1932, to June, 1934, this lease produced twenty cars of ore that contained about 0.71 ounces of gold and 3.5 ounces of silver per ton.[237] About fifteen men were employed.

Here, limestones of Carboniferous age strike north-northwestward and dip about 30° E. The underground workings, which consist of a 70-foot incline, a 70-foot drift, and an 87-foot inclined winze, indicate that the principal ore shoots plunge steeply northward along the bedding of the limestone. The largest ore body mined was about 70 feet long by 15 feet wide, but most of them were smaller, and all of them are very lenticular in form. The ore consists of fine-grained, vuggy, grayish to greenish-yellow quartz together with limonite, hematite, and a little wulfenite. Cerargyrite, locally associated with copper stain, is abundant in places. The gold is relatively fine grained and tends to occur where wulfenite is present.

MAMMOTH VICINITY

SITUATION AND HISTORY

The Mammoth district, sometimes considered part of the Old Hat district, centers about the Mammoth mine which is in Sec. 26, T. 8 S., R. 16 E. By road, it is 3 miles southwest of Mammoth,

---

235 Oral communication.
236 Oral communication from Mr. Poole.
237 Oral communication from Mr. Smith.

a town on the San Pedro River, and 21 miles south of Winkelman, the nearest railway station.

Prospecting was done in the Mammoth vicinity prior to the Civil War, but the modern locations of the Mammoth, Collins, and Mohawk properties were not made until about 1881.[238]

During the early eighties, C. R. Fletcher and associates blocked out a large deposit of gold ore in the Mammoth Mine, built a 30-stamp amalgamation mill at the San Pedro River, and established the town of Mammoth there. The Mammoth Gold Mines, Ltd., purchased the property in 1889, enlarged the mill to 50 stamps, and operated actively until 1893. The ore produced during 1890 averaged $14 per ton and cost $4 per ton for mining and milling.

The Collins property, next to the Mammoth on the west, was leased in 1894 by Johnston, Barnhart, and Collins, who, up to the end of 1895, mined more than 40,000 tons of ore from above the 300-foot level. This ore, which was treated in the Mammoth mill, probably yielded more than $240,000 worth of gold.

The Mammoth Gold Mining Company, successor to Mammoth Gold Mines, Ltd. acquired the Collins mine in 1898 and actively operated the Mammoth-Collins for four years. Tailings from the mill were treated in a 200-ton cyanide plant. From one hundred to a thousand men were employed during this period. Blake[239] states that, in April, 1901, a large stoping area at the north end of the mine suddenly caved from the surface to the 760-foot level and caused about an acre of the surface to subside for some 35 feet. Later, a drift around the caved section was driven on the 700-foot level.

The Mohawk Gold Mining Company acquired the Mohawk mine, adjoining the Mammoth on the east, in 1892 and operated a stamp mill during 1896 and 1897. This company sank a 500-foot shaft in 1906-1907, built a 30-ton mill, and operated until the end of 1912.

The Great Western Copper Company leased the Mammoth mine in 1913 and did considerable development work. In 1917, the St. Anthony Mining and Development Company leased the property for the purpose of working old tailings and stope-fills but ceased operations at the end of 1918.

In 1926, the New Year Group, adjoining the Mohawk on the east, was optioned by Sam Houghton who did some underground development and, in 1932, built a small mill. In 1933, a subsidiary of the Molybdenum Corporation of America obtained control of the New Year and Mohawk properties, built a cyanide plant, and carried on underground development. Towards the end of the year, about thirty-two men were employed.

---

[238] History abstracted in part from unpublished notes of J. B. Tenney and in part from material furnished by L. E. Reber, Jr., and P. C. Benedict, through the courtesy of the United Verde Copper Company.

[239] Blake, Wm. P., in Rept. of Gov. of Ariz., 1901, p. 189.

The total production of the Mammoth region has amounted to about $3,000,000, chiefly in gold of which the greater part has come from the Mammoth and Collins mines.

TOPOGRAPHY AND GEOLOGY

The Mammoth district is on a hilly, dissected pediment, at an altitude of about 3,200 feet, or 900 feet above the San Pedro River.

The rocks in this vicinity consist of coarse-grained granite, presumably of pre-Cambrian age; Mesozoic or Tertiary breccia and conglomerate, largely of andesitic to dacitic material; rhyolite, in part at least intrusive, probably of Tertiary age; and Pliocene Gila conglomerate. All of these formations have been subjected to rather intense faulting, the complex relationships of which are not thoroughly understood.

VEINS[240]

The veins of the Mammoth district occur within shear zones which strike west-northwestward and dip steeply southwestward. Their gangue consists of brecciated country rock, cemented and replaced with quartz and calcite together with some barite and fluorite. The quartz forms successive bands of which some of the latest generations show well-developed comb structure and are locally amethystine. Most of the gold occurs in an earlier generation of dense greenish-yellow quartz. The gold tends to be fine grained, but in places its particles are visible. In the oxidized zone are limonite, hematite, anglesite, cerussite, linarite, brochantite, malachite, azurite, calamine, wulfenite, vanadinite, pyromorphite, descloizite, cuprodescloizite, and pyrolusite. Galena, sphalerite, and sparse molybdenite occur as hypogene minerals.

Wall-rock alteration along the veins consists mainly of chloritization and silicification. Such alteration, together with the texture and mineralogy of the veins, points to deposition in the epithermal zone. No notable placers have been formed.

Ore bodies: The ore bodies of the Mammoth district occur in two roughly parallel groups. From northwest to southeast in the northern group are the Mammoth, Mohawk, and New Year properties, all of which appear to be on one vein, termed the Mammoth vein. In the southern group are the Collins property, on the Collins vein, and the Smith property, on the projected strike of the Collins vein. These veins dip steeply southwestward. With approximately the same strike but steep northeastward dip, the Mammoth fault comes to the surface near the Collins vein and, southeast of the Collins adit, transects its out-

---

[240] Much of the following information is abstracted from material furnished by L. E. Reber, Jr., and P. C. Benedict, through the courtesy of the United Verde Copper Company.

crop. The Mammoth vein is possibly a segment of the Collins vein that has been displaced for about 1,000 feet by the Mammoth fault. According to Reber and Benedict, this fault has been mineralized to some extent.

In the Mammoth mine, oxidation is complete to the 700-foot level. Water stands in the shaft at about 725 feet, but the ore on the 760-foot level is reported to be largely oxidized, with some residual bunches of galena. This galena is reported to contain 25 ounces of silver and 0.375 ounces of gold per ton. The shaft is stated to be 840 feet deep and to make 300 gallons of water per minute. Apparently there were two steeply southeastward-pitching ore shoots that diverged, branched, and became smaller upward. Between the 400-foot and 760-foot levels, these shoots ranged from 45 to 150 feet apart, from 7 to 40 feet in width, and from $6.24 to $11.96 in gold per ton. The northwestern shoot ranged from 200 to 395 feet in length and the southeastern from 90 to 220 feet. At 1934 prices, according to F. S. Naething, Vice President of the St. Anthony Mining and Development Company, the material between these ore shoots may constitute ore.

Blake[241], in 1901, stated that the stopes from the 200-foot to the 500-foot level of the Mammoth were in places 60 feet wide, and the ore probably averaged from $7 to $9 per ton. He also stated that, in 1898, 12 tons of wulfenite were shipped. By the end of 1900, the ore in the Mammoth mine above the 700-foot level had been largely stoped out. Some 10,000 tons that had been mined between the 700-foot and 760-foot levels averaged $8.63 per ton. The average costs per ton in 1900 were reported to be as follows: mining, $1.15; milling, $0.64; transportation by 3 miles of aerial tramway, $0.13; management, etc., $0.28; total $2.20. Previous to 1901, the recovery of gold by amalgamation ranged from 45 to 60 per cent, and cyanidation recovered from 50 to 65 per cent of the gold in the tailings.

An engineer who investigated the Mammoth mine particularly for wulfenite reported, in 1917, that there was no wulfenite above the 200-foot level; that the vein was ill-defined and of low grade on this level, but improved between the 200-foot and 300-foot levels; that, on the 600-foot level, the stope was 15 feet wide; and that a good grade of ore was present on the 700-foot level. He estimated that the ore averaged one per cent of MoO_3 and contained considerable vanadinite. F. S. Naething[242], however, states that wulfenite and vanadinite occur above the 200-foot level.

In the Mohawk mine, oxidation is complete to the deepest workings which are some 500 feet below the surface. The vein appears to be weak in the southeastern workings.

---

[241] Work cited.
[242] Oral communication.

The New Year workings, a short distance farther southeast, have crosscut a wide, low-grade vein on the 425-foot level. Some higher grade material has also been encountered. The outcrops to the southeast are mantled by Gila conglomerate.

The T. M. Smith property has been opened by a few hundred feet of tunnels, winzes, and raises. Here, cellular to cavernous brecciated zones in rhyolite contain thin, irregular gold-bearing quartz veins. The quartz is commonly made up of successive bands of fine-grained greenish-yellow, coarse dark-gray, coarse white, and coarse amethystine crystals with well-developed comb structure. The gold occurs mainly in the greenish-yellow quartz. In December, 1933, the property was under lease to A Aguayo and associates who had mined about three carloads of sorted ore.

In the Collins mine, oxidation is rather complete to the 650-foot level. A report by T. J. Davey in 1900 stated that the ore body between the tunnel level and the 226-foot level ranged from 110 to 160 feet long by 5 to 30 feet wide. On the 226-foot level, it averaged $6.16 per ton. Later, the Collins vein was explored from the 700-foot level of the Mammoth. On the 700-foot level of the Collins, about 80 per cent of the ore is primary, and 20 per cent partly oxidized. There have been silled out on this level some 2,578 square feet of material that averages 9.3 per cent of lead, 7.2 per cent of zinc, 0.5 per cent of copper, 1.7 ounces of silver, and 0.015 ounces of gold per ton, over an average width of 5.2 feet. Two diamond drill holes below this level have shown the vein to be of similar grade and width. Nearer the surface and southeast of the Collins adit, the Collins vein is capped by the Mammoth fault and has not been explored.

The Mammoth fault, which some geologists regard as the Dream vein, has been cut on the 500-foot level of the Mohawk and at a depth of 185 feet in the Brady workings, southeast of the New Year shaft. So far as explored, it contains only oxidized minerals.

FUTURE POSSIBILITIES

The future of the Mammoth district depends upon such factors as favorable metal markets, effective metallurgy and further exploration.

Wulfenite and vanadinite are sufficiently abundant in much of the oxidized ore to contribute materially to its value under favorable conditions.

Exploration of the oxidized zone of the Collins vein beneath the Mammoth fault and southeast of the Collins adit possibly would show gold-molybdenum ore.

Encouraging mineralization is reported to occur in veins which continue northwestward from the camp.

CASA GRANDE DISTRICT

MAMMON MINE[245]

The Mammon mine, near the western foot of the Slate Mountains, is accessible by one mile of road that branches from the Casa Grande-Covered Wells highway at a point about 32 miles from Casa Grande.

This deposit was discovered in 1892 by Morand and Peterson. During the following year, a company was organized, a well was sunk, and a 20-stamp amalgamation mill was built. Intermittent operations, carried on until the middle of 1897, yielded about $35,000 worth of gold from 2,500 tons of ore. After 1897, little work was done on the property until late in 1932 when surface trenching near the old workings revealed a parallel vein from which a test shipment of 2 tons returned more than $30 per ton in gold.

The mine is in a shallow basin floored essentially with pre-Cambrian schistose, sandy slates that strike N. 20° W., and dip 50° N. E. These slates are cut by numerous veins of white quartz that is locally stained with limonite, chrysocolla, and malachite and contains erratically-distributed free gold. The principal vein, where stoped, strikes N. 60° E. and dips 70° NW. but, some 200 feet northeast of the shaft, its strike changes to N. 20° W. In thickness, this vein ranges from a few inches to 3 feet and averages about 20 inches. It was opened by a 270-foot inclined shaft connected with several hundred feet of drifts, raises, and stopes. The stoping extends on both sides of the shaft to a depth of 250 feet and for a length of about 80 feet, on an ore shoot that was from one to 3 feet wide.

About 20 feet farther southeast, recent trenching exposed a parallel vein that averages about 20 inches wide for a length of 50 feet. It has yielded a little high-grade ore.

CHAPTER VIII—PIMA COUNTY

Pima County, as shown by Figure 10 (page 176), comprises an irregular area about 168 miles long by 74 miles wide. It consists of broad desert plains with scattered, north-northwestward-trending mountain ranges of pre-Cambrian schist and granite, Paleozoic and Cretaceous sedimentary beds, Cretaceous and Tertiary granite and porphyry, and Tertiary volcanic rocks.

This county, which ranks seventh among the gold-producing counties of Arizona, to the end of 1931, produced approximately $3,212,000 worth of gold of which about $2,050,000 worth was a by-product from copper ores, $16,000 a by-product from lead ores.

---

[243] Abstracted from unpublished notes of J. B. Tenney, 1933.

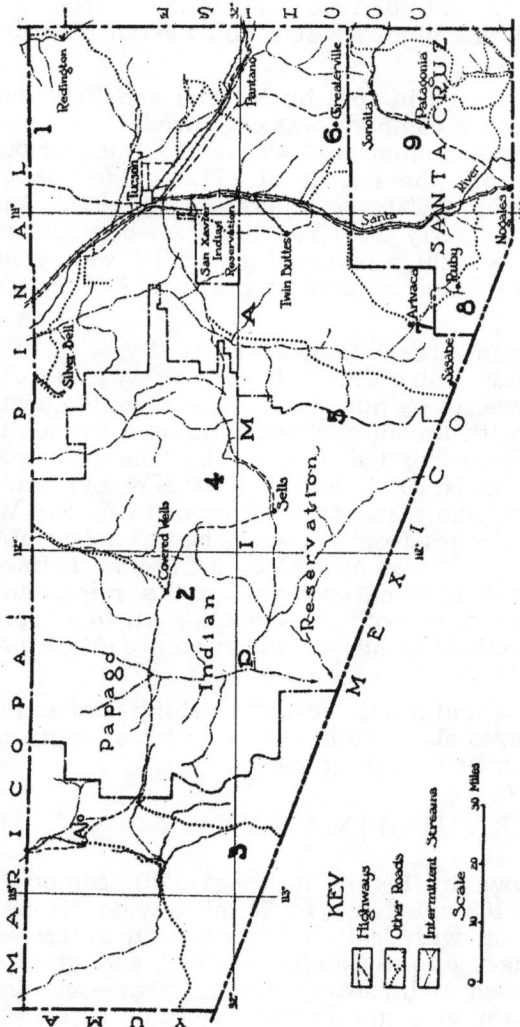

Figure 10.—Map showing location of lode gold districts in Pima and Santa Cruz Counties.

PIMA COUNTY DISTRICTS

1 Old Hat
2 Quijotoa
3 Puerto Blanco Mountains
4 Comobabi

5 Baboquivari
6 Greaterville
7 Arivaca

SANTA CRUZ COUNTY DISTRICTS

8 Oro Blanco

9 Wrightson

$1,050,000 was from placers, and $96,000 was from lode gold mines.[244]

CABABI OR COMOBABI DISTRICT

The Cababi or Comobabi district, of central Pima County, is in the South Comobabi Mountains and within the Papago Indian Reservation.

The southern portion of these mountains is a faulted complex of steeply dipping sedimentary beds together with abundant dikes and stocks of dioritic to basic porphyrys. The sedimentary rocks are largely massive, impure shales which strongly resemble certain Cretaceous formations of southern Arizona.

A small production of copper, lead, and silver ores has been made in the western portion of the range, but gold-bearing quartz veins are found principally in the southeastern and southern margins. These veins dip steeply, strike in various directions, and are traceable for considerable lengths on the surface. They commonly range in width from less than one inch to 4 feet. Their quartz, which is associated with more or less calcite, generally appears in at least two generations of which the later is coarsely crystalline and the earlier is of massive, sugary to dense texture. The gold occurs mainly with particles and bunches of iron minerals within the sugary quartz. In the oxidized zone, which is of irregular depth, the gold is associated with black iron oxides, but, below this zone, it is contained in pyrite. The principal wall-rock alteration is to quartz and sericite.

AKRON MINE

The Akron property, formerly known as the Jaeger, consists of twenty claims at the southeastern base of the South Comobabi Mountains, about 4 miles west from the Ajo highway and 54 miles from Tucson.

This ground was obtained several years ago by G. H. Jaeger who shipped a few cars of high grade gold ore from the main shaft workings. Later, a few shipments were made by G. Williams, a lessee. The Akron Gold Mining Company, which acquired the property in 1932, reports a production of about twelve cars of ore up to the end of December, 1933. At the time the property was visited, this company was carrying on considerable prospecting and had started the construction of a 100-ton cyanide and amalgamation mill.

In this vicinity, an undulating pediment exposes small areas of shales intruded by dioritic to rather basic dikes which strike in various directions and have been extensively faulted. Several of the fault fissures contain abundant limonite, together with thin, lenticular, quartz-carbonate veins of the type described.

---

[244] Statistics compiled by J. B. Tenney.

At the collar of the old main shaft, several narrow quartz veins form a lode, from one to 3 feet wide, that strikes eastward and dips nearly vertically. This shaft is reported to be 280 feet deep and to connect with some 800 feet of workings. In December, 1933, these workings were full of water to within 80 feet of the surface.

CORONA GROUP

The Corona group of 7½ claims, held by T. P. Quinn and associates, is 1½ miles northwest of the Akron mine. The principal vein strikes northwestward, dips 60° SW., and is traceable southeastward for more than half a mile. In December, 1933, the principal opening on this vein was an 82-foot inclined shaft. Here, the footwall is diorite porphyry and the hanging wall is slate. The vein, which is about 4 feet wide, consists of dense bluish-white quartz with abundant dark inclusions and scattered bunches of pyrite. A 1½ ton test shipment of this ore is reported to have carried about an ounce of gold per ton.

OTHER CLAIMS

Several small lots of rich gold ore have been mined by Gus H. Jaeger from the Hawkview claims, about 2 miles from the Akron camp. This ore was treated in a small amalgamation mill.

The Faro Bank group of claims, held by M. M. Holmes, is at the southern edge of the range, about 4 miles north of Sells. During the past few years, several small shipments of gold ore have been made from this property.

QUIJOTOA MOUNTAINS

The Quijotoa Mountains, in the Papago Indian Reservation, south-central Pima County, contain silver and gold deposits whose total yield amounts to nearly $500,000. Although most of this output has been in silver, the gold deposits have attracted considerable local attention.

This range is about 15 miles long by a maximum of 5 miles wide and rises to approximately 4,000 feet above sea level or 1,500 feet above the plain. It is formed principally of quartz monzonite, minor sedimentary beds, thick andesitic flows, stocklike masses of quartz diorite, and minor dikes. These rocks have been considerably affected by faulting principally of northwestward trend.

This region is hot in summer. Water is obtained from shallow wells and shafts in the pediment on both sides of the mountains.

Gold-bearing veins: Quartz veins and hematitic brecciated zones are numerous in the Quijotoa Range. Although some of them, particularly in the northern half of the area, have afforded spectacular gold specimens, they have yielded only a small production of gold ore.

The Morgan mine,[243] owned by the Larrymade Mines, Inc., is 6 miles south of Covered Wells. Its workings include a 300-foot inclined shaft, a few hundred feet of drifts, and several shallow openings, mostly in a shear zone that strikes northwest and dips 65° NE. The vein filling consists chiefly of banded, coarse, dense, grayish-white quartz, commonly accompanied by dark-red gouge. This quartz forms lenticular masses, from 10 to 30 feet long by one to 4 feet wide, which fray out abruptly into stringers. It is accompanied by calcite, iron oxides, and some manganese dioxide. The ore consists mainly of dense, gray, brecciated quartz cemented in part with hematite and calcite. The gold occurs chiefly in this cementing material and is in places associated with manganese dioxide. The principal streak of relatively high-grade ore was from 6 inches to 2 feet wide. The quartz diorite wall rock has been somewhat altered to calcite and sericite.

One carload of ore, shipped in 1931 from the upper part of this mine, contained 1.39 ounces of gold and 1.11 ounces of silver per ton. Three carloads of similar ore were shipped in 1932.

BABOQUIVARI MOUNTAINS

The Baboquivari Mountains extend for 30 miles northward from the international boundary. For most of this distance, their crest line marks the eastern border of the Papago Indian Reservation.

The middle segment of the range consists mainly of metamorphosed Cretaceous strata which have been intruded by numerous dikes and complexly faulted. Northward, these formations give way to granite and gneiss. The sedimentary rocks tend to form rounded slopes, while the granite outcrops are rugged, and the dike exposures commonly stand out as cliffs and crags. As Bryan states,[246] well-developed pediments extend into the mountains in deep coves and reentrants along the courses of the major mountain canyons.

The principal gold-bearing quartz veins known in these mountains occur northwest and southeast of Baboquivari Peak. Their total production, which amounts to about $142,000, has come mainly from the Allison mine.

WESTERN PORTION

Considerable gold prospecting has been carried on in the western or Papago Indian portion of the Baboquivari Mountains. The only known production from this part of the range has come from the Allison or Chance mine which is accessible from Sells by 21 miles of road.

---

[243] Description abstracted from unpublished notes of G. M. Butler and also from Gebhardt, R. C., Geology and mineral resources of the Quijotoa Mountains: Unpublished M. S. Thesis, Univ. of Arizona, 1931

[246] Bryan, Kirk, U. S. Geol. Survey Water-Supply Paper 499, pp. 247-48 1925.

The first record[247] of mining activity here was in 1898 when Wm. P. Blake reported that the Allison mine had been developed by a 100-foot shaft. He reported that a small production of rich, sorted gold ore was made. This rich ore proved to be rather superficial, and no further work was done until 1923 when a tunnel was driven under the old shaft. During 1924 and 1925, some gold-silver bullion and concentrates were produced. The Tom Reed Gold Mines Company leased the property in 1926, did considerable underground work, and built a small flotation mill. Due to its high content of manganese oxides, the ore was rather refractory. Production from 1926 to 1928, inclusive, amounted to 2,176 ounces of gold and 44,705 ounces of silver. In 1930, the El Oro Mining and Milling Company obtained control of the property. About $5,500 worth of gold-silver bullion and a few tons of concentrates were produced in 1931. The mine has been idle since that year.

Here, tilted conglomerate beds, intruded by dikes of rhyolite and andesite porphyry, form a broad foothill belt. These rocks contain numerous thin stringers of gold-bearing quartz, locally associated with pyrite, which have been prospected to some extent.

The Allison vein strikes eastward and is about 30 feet in maximum width. Its quartz is grayish white to olive gray in color. The vein as a whole is of rather low grade, but certain portions, containing abundant manganese and iron oxides, were relatively rich in gold and silver. Workings on this property include a 320-foot adit, a 625-foot inclined shaft, and considerable drifting. Most of the ore mined during recent years is reported to have occurred below the 400-foot level.

SOUTHERN PORTION

In the southern portion of the Baboquivari Mountains most of the gold-quartz veins so far explored occur near the foothill margin of a pediment along the eastern base of the range. Here, the prevailing rocks consist of unmetamorphosed to schistose Cretaceous shales, conglomerates, and quartzites, intruded by numerous dikes of granite and diorite porphyry. The veins occur within fault fissures which strike in various directions and dip from 15° to nearly vertical. Their filling consists of coarse-grained white quartz and calcite, together with scattered bunches of pyrite, galena, and minor chalcopyrite. In places, molybdenite is relatively abundant. Within the oxidized zone, which is only a few feet below the surface, the quartz is rather cellular and contains abundant limonite with some malachite. Sericite and secondary quartz are abundantly developed in the wall rocks.

The gold of these veins is generally not visible, but it probably occurs in the sulphides and in their oxidation products.

---

[247] Unpublished historical notes of J. B. Tenney.

Operations: During the eighties, according to O. C. Lamp,[248] a 270-foot shaft was sunk on the old Gold Bullion claim. About 3,000 tons of ore was mined from its vein and put through a stamp mill.

Within the past few years, considerable prospecting has been carried on in the area. At the Iowana property of twenty-six claims, which is 6 miles west of the Sasabe road and 62 miles from Tucson, a few shipments of table and flotation concentrates have been made. According to O. C. Lamp, President of the company, these concentrates contained from 4 to 7 ounces of gold and 58 to 60 ounces of silver per ton.

On the Jupiter property in the same vicinity, several hundred feet of workings were run and some ore was produced. A little underground exploration was done by the Shattuck-Denn Mining Corporation.

In December, 1933, eight men were working at the Gold King property, 2 miles southwest of the Iowana. According to F. R. Mooney, one of the operators, several tons of sorted ore, containing from $40 to $60 in gold and silver per ton, had been shipped.

ECONOMIC POSSIBILITIES

In the Baboquivari Mountains, vein outcrops tend to be inconspicuous and in many places are more or less concealed by faulting and by talus or vegetation. Geophysical prospecting might be helpful in the southeastern portion of the range where the gold is associated with sulphides which are not completely oxidized beyond shallow depths. The most favorable areas appear to be the pediments and the immediately adjacent foothills.

Due to the intensity of post-mineral faulting in the range, many of the veins may be followed by mining only after a thoroughly comprehensive geologic study of the area has been made.

PUERTO BLANCO MOUNTAINS

Gold-bearing quartz veins occurring about 6 miles west of Dripping Spring in the Puerto Blanco Mountains have been prospected by a few shallow shafts on the Golden Bell claims. This area, which is about 30 miles south of Ajo, via a branch from the old Sonoita road, forms part of a schist pediment at the western foot of the range.

The vein exposures, which are more or less interrupted by later gravels, appear along an eastward-trending belt about a mile in length. At the western margin, the vein is vertical and consists of dense shiny white quartz, somewhat iron-stained and brecciated near the walls. About a mile farther east, the exposures are from one to 2 feet wide and have been somewhat offset by cross-faulting. Here, a shallow shaft shows shiny,

---

[248] Oral communication.

grayish-white to greenish-yellow quartz with some masses of cellular limonite. The schist walls of the veins are considerably sericitized and silicified.

According to Charles E. Bell, one of the owners of these claims, the exposed eastern portion of the vein in places averages about 0.5 ounce of gold per ton.

CHAPTER IX—GILA COUNTY

Gila County, as shown by Figure 9, (page 159), comprises an irregular area about 115 miles long by 70 miles wide. Its western half consists of a series of northwestward-trending mountain ranges of complexly faulted pre-Cambrian metamorphic, igneous, and sedimentary rocks, Paleozoic sedimentary beds, and Tertiary volcanic rocks. In the southwestern portion of the area, west of Miami, is the Shulze granite stock that probably gave rise to the Globe and Miami copper deposits.[249] The eastern portion of the county consists essentially of a dissected plateau of pre-Cambrian, Paleozoic, and Tertiary rocks that, so far as known, contain no commercially important gold deposits.

Gila County, which ranks eighth among the gold-producing counties of Arizona, has yielded approximately $3,100,000 worth of gold of which about $2,800,000 worth has been a by-product of copper mining.[250]

As indicated on Figure 9, lode gold deposits occur in the Payson, Banner, Globe, and Spring Creek districts. The Banner, Globe, and Spring Creek deposits together have yielded a few tens of thousands of dollars' worth of gold ore, but the field work for the present report had to be terminated before they could be studied. Some data on the Banner district are given by Ransome[251] and by Ross.

PAYSON DISTRICT[252]

Situation and accessibility: Payson, a small resort and cattle town in northern Gila County, is 75 miles by road from the railway at Clarkdale and 90 miles from Globe and Miami.

History: The earliest mineral locations in this vicinity were made in 1875. Within three years, practically all the prominent quartz veins were located, and, by 1881, more than three hundred men were attracted to the district. The cream of production was

---

[249] Ransome, F. L., Copper Deposits of Ray and Miami, Arizona: U. S. Geol. Survey Prof. Paper 115, 1919.

[250] Statistics compiled by J. B. Tenney.

[251] Ransome, F. L., Description of the Ray Quadrangle: U. S Geol. Survey Folio 217, p. 23, 1923.
    Ross, C. P., Ore deposits of the Saddle Mountain and Banner Mining Districts, Ariz.: U. S. Geol. Survey Bull. 771, 1925.

[252] Lausen, Carl, and Wilson, Eldred D., Gold and copper deposits near Payson, Arizona: Ariz. Bureau of Mines Bull. 120.

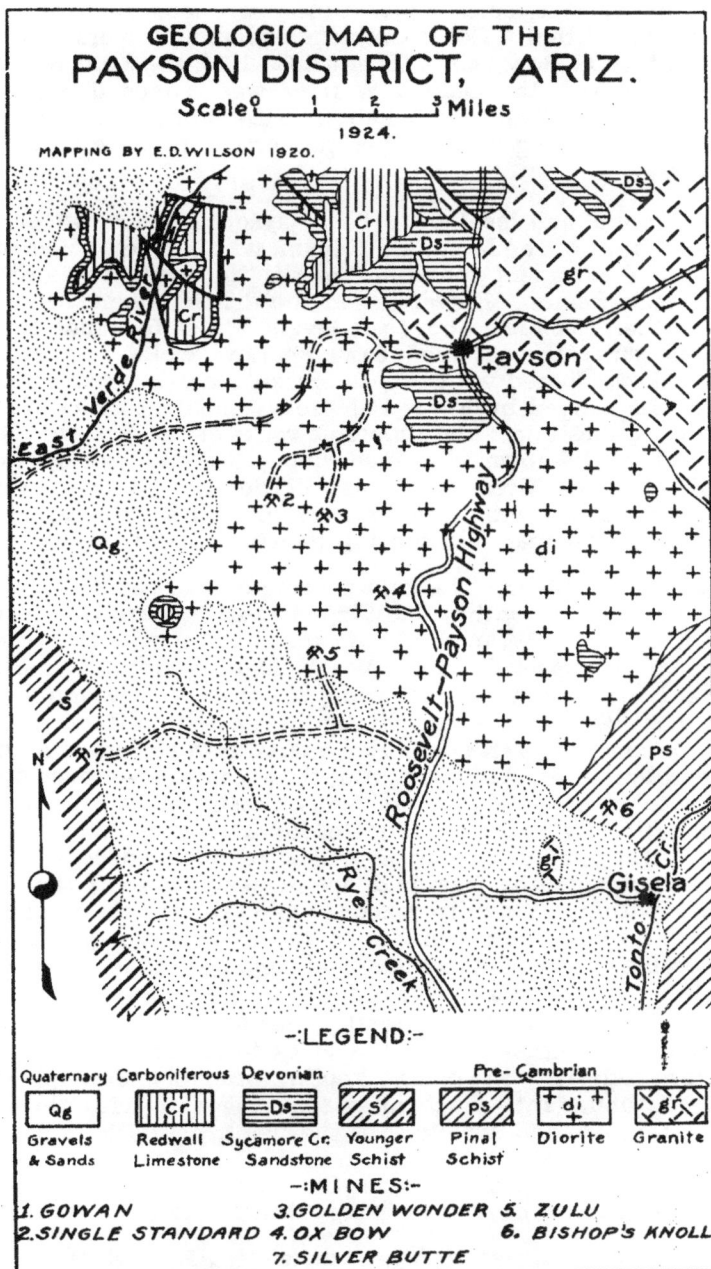

GEOLOGIC MAP OF THE
PAYSON DISTRICT, ARIZ.
Scale 0 1 2 3 Miles
1924.
MAPPING BY E.D.WILSON 1920.

-:LEGEND:-

Quaternary Carboniferous Devonian — — — Pre-Cambrian

Qg	Cr	Ds	S	Ps	di	gr
Gravels & Sands	Redwall Limestone	Sycamore Cr. Sandstone	Younger Schist	Pinal Schist	Diorite	Granite

-:MINES:-

1. GOWAN   3. GOLDEN WONDER   5. ZULU
2. SINGLE STANDARD   4. OX BOW   6. BISHOP'S KNOLL
          7. SILVER BUTTE

Figure 11.

harvested by 1886, since which time operations have been generally intermittent. Considerable development work has been done on the Ox Bow vein   During 1933-1934, the Zulu was being actively worked, the Golden Wonder was reopened, work was done on the Wilbanks-Callahan and De Ford properties, and plans for reopening a few other properties in the district were under way.

Topography and geology: The Payson district is near the northern limit of the Mountain Region, within a few miles of the Mogollon escarpment that here marks the southern border of the Plateau Region. Elevations in the district range from 3,400 to more than 5,000 feet above sea level. Water for domestic purposes is generally obtained from shallow wells. At Payson, broad valleys floored with pre-Cambrian granitic rocks alternate with mesas capped by lower Paleozoic sandstone which, farther north, continues under the Paleozoic rocks of the Plateau. South and west of the town, dissected slopes descend steeply to the valleys of Tonto Creek and East Verde River. These slopes, which are floored with a faulted complex of schist and diorite, contain the principal gold-bearing quartz veins of the district.

Gold-bearing veins: The gold-bearing quartz veins of the Payson district occur within fault zones that generally range in strike from N. 15° W. to N. 65° W. and dip northeastward. Most of the veins, especially the Zulu, Golden Wonder, and Single Standard, are less than 2 feet wide, but the Ox Bow and Gowan attain considerably greater widths. The oxidized portions of the veins consist of rather cellular quartz with considerable hematite and limonite. At least part of these cavities were originally filled with pyrite. Where the veins consist of rather massive quartz with only a small amount of hematite and limonite, they are of lower grade. Locally, bunches of oxidized copper minerals, accompanied by free gold, are present. Most of the workings are within the oxidized zone. Typical vein material from below the water table consists of rather massive quartz with considerable pyrite and a little chalcopyrite. It is reported to carry less than an ounce of gold per ton.

The wall rock, which is generally diorite, shows alteration for several feet on either side of the veins to chlorite, sericite, and secondary quartz. In places kaolin, probably derived from sericite, is abundant. Such alteration, together with the mineralogy, texture, and structure of the veins, points to deposition in the mesothermal zone.

For descriptions of the Gowan, Ox Bow, and Single Standard mines, see Arizona Bureau of Mines Bulletin 120.

Economic possibilities of Payson district: Although the quartz veins of the Payson district carry some gold and silver throughout, only certain portions of them are rich enough to constitute ore under present economic conditions. Because most of the old workings were inaccessible when the district was studied, most of the factors governing the occurrence of these ore shoots remain unknown. In general, the quartz veins are richest where they widen or where they carry notable amounts of oxidized iron or copper minerals. Below water level, the gold is probably contained in pyrite.

CHAPTER X—GREENLEE COUNTY

Greenlee County, as shown by Figure 12, (page 186), comprises an irregular area about 93 miles long by 26 miles wide. It consists largely of rugged mountains of Tertiary volcanic rocks that surround the Clifton-Morenci area of pre-Cambrian schist and granite, Paleozoic and Cretaceous sedimentary beds, and Cretaceous or Tertiary porphyry.

This county, which ranks ninth among the gold-producing counties of Arizona, to the end of 1931, produced approximately $1,762,000 worth of gold, most of which was a by-product from copper ores from the Morenci district.[253]

MORENCI REGION

The copper ores of the Morenci region contain very little gold, but, as Lindgren[254] has shown, the outlying deposits of the same general character and age contain less copper and more gold.

In Gold Gulch, west of Morenci, gold-bearing veins have been worked intermittently and on a small scale for many years. Lindgren says: "The diorite-porphyry here contains many included masses of limestone and other sediments. Many narrow and irregular veins cut these rocks, and pockets of gold associated with limonite have been found in several places. The veins are small, no great depth has been attained by the workings, and the deposits, which farther down will doubtless contain sulphide ore, have not yet proved to be of much value."

Lakemen mine:[255] The Lakemen property of eight patented claims, held in 1934 by Mrs. Grace Morrison, is in Gold Gulch, about 4½ miles southwest of Morenci. Between 1885 and 1895, according to local reports, the property produced notable amounts of gold ore. Its underground workings include a 300-foot vertical shaft with about 470 feet of drifts and some small stopes on the 200- and 300-foot levels. In 1934, these workings were reopened, and some shipping ore was mined.

The vein strikes N. 60° E., dips vertically, and occurs in diorite-porphyry. Its width ranges up to 10 feet and averages about

[253] Statistics by J. B. Tenney.

[254] Lindgren, W., Copper deposits of Clifton-Morenci district, Arizona: U. S. Geol. Survey Prof. Paper 43, p. 211, 1905; Clifton Folio (No. 129), p. 13, 1905.

[255] Description based on notes supplied by R. J. Leonard, 1934.

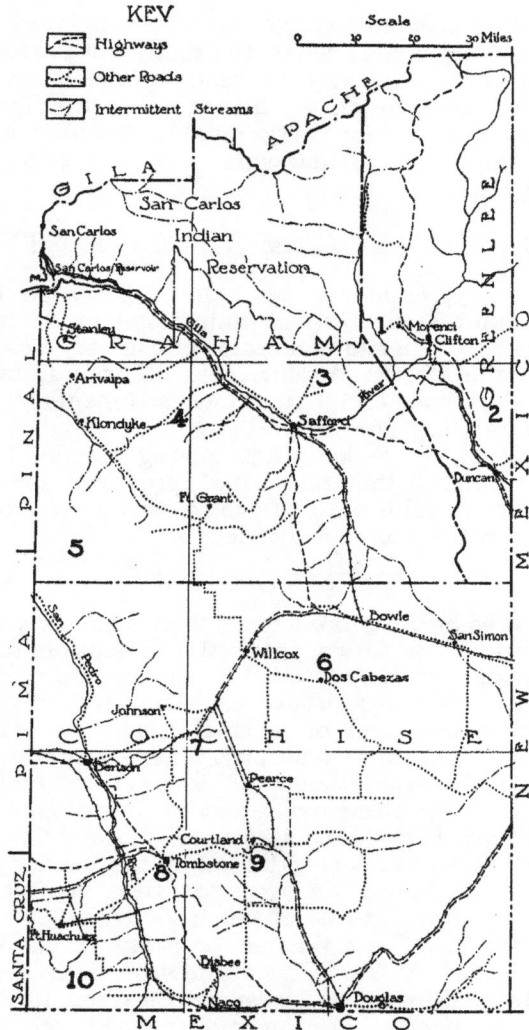

Figure 12.—Map showing location of lode gold districts in Greenlee, Graham, and Cochise Counties.

GREENLEE COUNTY DISTRICTS

1 Gold Gulch (Morenci)   2 Twin Peaks

GRAHAM COUNTY DISTRICTS

3 Lone Star   5 Rattlesnake
4 Clark

COCHISE COUNTY DISTRICTS

6 Dos Cabezas, Teviston   9 Turquoise
7 Golden Rule   10 Huachuca
8 Tombstone

3 feet. Its gangue consists of brecciated, coarse-textured, grayish quartz and sericitized porphyry, cemented by a later generation of vuggy quartz. Limonite stain is abundant in places, and some specimens of the ore show finely disseminated pyrite. The diorite-porphyry wall rock is somewhat sericitized, silicified, and iron stained. The mine makes about 1,200 gallons of water per day.

Hormeyer mine: The Hormeyer mine,[256] about one mile east-southeast of Morenci, is believed to have yielded $30,000 worth of gold-lead-copper ore prior to 1902. The vein, which strikes northeastward and follows a porphyry dike in Ordovician limestone, forms an outcrop of cellular quartz stained yellow by lead oxide. Its ore contains native gold, abundant lead carbonate, and a little copper. The underground workings include two tunnels. The lower tunnel, which was run along a porphyry dike 6 feet wide, found no ore.

Copper King Mountain vicinity: Of the Copper King Mountain area, some 2 miles east of Metcalf, Lindgren[257] says: "The ore of Copper King mine contains from $1 to $4 per ton in gold. Northeast of Copper King Mountain the same vein system continues in granite, usually following porphyry dikes, but here carries less copper and considerably more gold. The croppings yield light gold in the pan, and, in tunnels 50 to 100 feet below, sulphide ore is found in many places, consisting of auriferous pyrite, chalcopyrite, zinc blende, and galena. The value of these veins is as yet problematical."

CHAPTER XI—SANTA CRUZ COUNTY

Santa Cruz County, as shown by Figure 10, (page 176), comprises an irregular area about 53 miles long by 27 miles wide. It consists of wide plains and large mountain ranges of Paleozoic and Cretaceous sedimentary beds, Cretaceous and Tertiary granite and porphyry, and Tertiary volcanic rocks.

This county, which ranks tenth among the gold-producing counties of Arizona, to the end of 1931, yielded approximately $1,279,000 worth of gold of which nearly $200,000 worth was a by-product from copper and lead ores.[258]

ORO BLANCO DISTRICT

Situation: The Oro Blanco district is in the southeastern portion of the Oro Blanco Mountains, of southwestern Santa Cruz County. Its principal settlement, Ruby, in the eastern part of the district, is 32 miles by road from Nogales and about 34 miles

---

[256] Description abstracted from Lindgren, W., U. S. Geol. Survey Prof. Paper 43, p. 212, 1905.

[257] Work cited, p. 212.

[258] Statistics compiled by J. B. Tenney.

from Amado, a station on the Nogales branch of the Southern Pacific Railway.

History and production:[259] Some of the gold deposits in the Oro Blanco district were probably worked in a small way by the early Spanish explorers. The first American locations were made in 1873 on the Oro Blanco vein. The Yellow Jacket, Ostrich, and other locations were made soon afterward, and the richer ore was treated in arrastres. The Ostrich mill, equipped with a roasting furnace to treat refractory sulphide ores, was built during the early eighties and operated by the Orion Company on ore from the Montana and Warsaw mines. The Warsaw mill was built in 1882 and operated as a customs plant. In 1884, the Esperanza Mines Company built a mill to treat ore from the Indestroth or Blain ledge.

From 1887 to 1893, most of the mines were inactive. By 1894, however, mills were operating at the Austerlitz and Yellow Jacket mines, mills were being built at the Montana and Old Glory mines, and the Ragnarole, Golden Eagle, St. Patrick, Tres Amigos, San Juan, Franklin, Cleveland, Oro, Nil Desperandum, and Last Chance deposits were being developed. In 1896, small mills were built at the Oro and Golden Eagle mines. In 1903, amalgamation and cyanide mills were built to treat ores from the Golden Eagle, Oro Blanco, Tres Amigos, and Ragnarole mines.

Lode Gold Production, Oro Blanco District:[260]

1873 - 1886 ....$ 700,000	Oro Blanco, Ostrich, Yellow Jacket, Warsaw, and Montana. (Estimated).
1894 - 1896 300,000	Austerlitz, Montana, Old Glory, Ragnarole, Golden Eagle, Tres Amigos. (Estimated).
1903 7,500	Old Glory. (Estimated).
1904 30,000	Oro Blanco, Golden Eagle, Ragnarole, Sorrel Top, and Tres Amigos. (Estimated).
1907 1,000	
1909 1,523	
1912 7,555	
1913 12,704	
1914 2,485	
1915 6,100	
1916 6,057	
1921 300	Estimated
1922 1,000	Estimated.
1923 1,000	Estimated.
1928 15,188	Montana Mine
1929 36,115	Montana Mine
1932 3,315	
Total $1,131,842	

[259] Largely abstracted from unpublished notes of J. B. Tenney.
[260] Figures compiled by J. B. Tenney.

After 1904, little work was done in the district until 1912 when the Austerlitz and Oro mines were reopened. A concentrator was built at Austerlitz and operated for slightly more than one year.

Between 1914 and 1931, the gold mines of the district were practically idle. During this period, the Montana mine, which, during the early days, produced gold ore from near the surface, was developed into an important lead-zinc-copper deposit.

Since 1931, some of the gold mines, notably the Tres Amigos and Margarita, have been reopened.

Topography and geology: The mineralized portion of the Oro Blanco Mountains consists of an uplifted table land, dissected into a series of ridges and southward-trending canyons that carry water during part of the year. The altitude ranges from 3,600 to 4,900 feet. Except in the canyons, vegetation is sparse.

The geology is rather complex and inadequately understood. The principal formation is a series of more or less metamorphosed arkosic sandstones, quartzites, conglomerates, and shales, with some intercalated volcanic rocks. These beds, which are probably of Cretaceous age, rest upon an irregular surface of altered coarse-grained, grayish diorite. They have been intruded by dikes of basic to acid composition and subjected to complex faulting. East of Oro Blanco Viejo Canyon, they are overlain by a thick succession of volcanic rocks.

Types of gold deposits: The gold deposits of the Oro Blanco district include three principal types: (1) Sulphide-bearing quartz veins, exemplified by the Old Glory and Austerlitz mines. (2) Mineralized shear zones, as at the Tres Amigos, Dos Amigos, and Oro Blanco mines, and (3) mineralized bodies of country rock, as at the Margarita mine.

OLD GLORY MINE

The Old Glory mine is west of the Warsaw road and about 2 miles west of Ruby, at an altitude of 4,600 feet. During the early nineties, this property was equipped with a 40-ton mill that operated, whenever water was available, until 1898. In 1902, a 30-stamp mill and a water-impounding dam were built, but operations ceased in 1903. According to M. J. Elsing,[261] present owner of the property, approximately 2,500 tons of ore that averaged $3 per ton was milled during 1902-1903.

Here, the prevailing rocks are metamorphosed grits of probable Cretaceous age, intruded by altered basic dikes that strike northeastward. The vein, which outcrops on top of the ridge west of the mill, strikes southeastward and dips steeply northeastward. On the surface, it is traceable for a distance of several hundred feet and, in places, is more than 50 feet wide. The gangue is coarse-textured, massive white quartz. In the ore shoots, ir-

---

[261] Oral communication.

regular disseminations and bunches of auriferous pyrite are present. Near the surface, most of the pyrite is oxidized to limonite and hematite.

Developments on the Old Glory property include several open cuts and about 500 feet of underground workings, mostly tunnels. One of the open cuts is approximately 100 feet long by 10 to 35 feet wide by 25 feet deep. According to Mr. Elsing,[261] the mine contains a large tonnage of material that averages about 0.137 ounces of gold and one ounce of silver per ton.

AUSTERLITZ MINE

The Austerlitz mine is on the west side of the Arivaca road, about 2¾ miles northwest of Ruby. During the nineties, this property was equipped with a mill and made a considerable production, but no records of the amount are available. In 1912, the mine was reopened and equipped with a concentrator that was operated for slightly more than a year.

Milton[262] describes the mine as follows: "This deposit is a flat vein 4 to 12 feet wide carrying gold and silver asociated with pyrite and chalcopyrite. The mine is credited with having produced $90,000 from a small ore shoot. It has been developed by about 800 feet of adit and one shaft 125 feet off the vein."

TRES AMIGOS, DOS AMIGOS, AND ORO BLANCO MINES

The Tres Amigos, Dos Amigos, and Oro Blanco mines are about 4 miles by rough road southwest of Ruby. Since early 1933, these mines have been held by Legend Group, Ltd., which has done a little development work and conducted tests with a 100-ton concentrating mill. Water was pumped from a depth of 150 feet in the Oro Blanco shaft.

The Tres Amigos mine, ¾ mile by road south of the mill, has been opened by an adit, reported to be 2,000 feet long, with four winzes of which two are said to be 100 feet deep, and several large stopes. Its ore occurs within a nearly vertical fault zone that strikes N. 30° W. For approximately 475 feet, the adit is in andesite that, northward, gives way to decomposed, sericitized diorite. The stopes, which are within the first 570 feet of the adit, range up to 35 feet in width by 60 feet in height. They are in irregular, lenticular bodies of brecciated to pulverized country rock. This material is considerably sericitized and locally contains abundant iron and manganese oxides. According to Keyes,[263] samples taken across this brecciated zone assayed from 0.1 to 1.0 ounce of gold per ton, but the material at permanent water level is practically barren.

---

[262] Milton, M. C., The Oro Blanco district of Arizona: Eng. and Min. Jour., vol. 96, p. 1006, 1913.

[263] Keyes, C. R., Tres Amigos gold veins of Arizona: Pan. Am. Geol., vol. 39, pp. 159-60, 1923.

The gold occurs as mediumly fine to coarse particles. The ore is said to contain about 10 ounces of silver per ounce of gold.

Southeast of the Tres Amigos adit, the brecciated zone has been opened by two shallow shafts and some lateral work in the Monarch mine. In May, 1934, a little high-grade ore from this mine was being milled in an arrastre.

The Oro Blanco and Dos Amigos mines are about ¾ mile north of the Tres Amigos workings. The mineralized shear zones here strike about N. 35° W. dip almost vertically, and are of the same type as that of the Tres Amigos mine. The following statements regarding the old Oro Blanco workings are supplied by officials of Legend Group, Ltd. The shaft is 265 feet deep and contains water to the 150-foot level. On the 125-foot level, a drift that extends northwestward for 1,500 feet contains several stopes that range up to 50 feet wide by 100 or more feet high. A 700-foot cross-cut southwestward cuts three other mineralized shear zones that have been explored by short drifts and found to contain large amounts of medium to low-grade ore.

The Dos Amigos mine, on the opposite side of the gulch, southeast of the Oro Blanco shaft, has been opened by a 100-foot shaft with some drifts and an adit tunnel more than 700 feet long that extends through the hill. Here, the principal rocks are complexly faulted, arkosic conglomerates. As seen in the adit tunnel, the fault zone contains irregular bodies, up to several feet wide, of brecciated to pulverized rock with considerable iron and manganese oxides.

MARGARITA MINE

The Margarita mine is north of the Old Glory mine and about 2 miles west of Ruby. During the nineties, when known as the McDonald prospect, it was opened by some 1,200 feet of tunnels and shallow workings from which a little gold was produced Since November, 1931, the Margarita Gold Mines Company has carried on a little development work, built a 50-ton cyanide mill, and produced two small lots of bullion as a result of testing operations. Water is piped from a tunnel in Old Glory Canyon, 1,960 feet distant.

In the vicinity of the mine, the prevailing rocks are metamorphosed, siliceous Cretaceous strata, intruded by dikes of dioritic and rhyolitic porphyry.

The workings immediately northwest of the mill have opened a mineralized zone about 200 feet long by 100 feet wide by 80 feet deep. This zone consists of silicified, sericitized country rock with locally abundant limonite and hematite. Its richest material contains thickly disseminated pseudomorphs of limonite after pyrite. According to officials of the Margarita Gold Mines Company, this zone contains considerable material that averages 0.3 ounce of gold and 0.5 ounce of silver per ton.

The workings east of the mill have opened a mineralized zone about 200 feet long by 30 feet wide by 50 feet deep. It is similar ın character to the zone already described, but carries pyrite in ıts lower portion.

CHAPTER XII—GRAHAM COUNTY

Graham County, as shown by Figure 12 (page 186), comprises an irregular area about 85 miles long by 72 miles wide. It consists of wide plains surmounted by large, northwestward-trending mountain ranges of pre-Cambrian schist and granite, Paleozoic and Cretaceous sedimentary beds, Cretaceous granite-porphyry, and Tertiary volcanic rocks.

This county has produced only a relatively small amount of gold.

LONE STAR DISTRICT

The Lone Star district, of Graham County, is in the southwestern portion of the Gila Mountains, northeast of Safford. It is accessible by unimproved desert roads from Pima, Safford, and Solomonsville.

Since 1906, this district has produced less than $75,000 worth of copper, lead, and silver, together with a small amount of gold. During the past few years, considerable prospecting for gold has been done by tunnels and shallow shafts in the mountains north of San Juan and Lone Star mines, on the Roper, West, Wickersham, Merrill, and other properties.

In this district, the southwestern slope of the Gila Mountains consists of andesitic lavas and flow breccias, intruded by dikes and masses of diorite-porphyry.

In the vicinity of Walnut Spring and north of the Lone Star mine, gold occurs as fine to visible flakes in certain fractured and sheeted zones within and near the diroite dikes. These zones strike eastward and northward, dip steeply, and are traceable for long distances on the surface. They contain scattered, irregular areas of iron and manganese oxides and a few thin, very discontinuous veins of medium to fine-grained quartz with local pseudomorphs of limonite after pyrite. In places, small amounts of copper stain are present. The gold occurs in wall-rock fractures associated with the iron and manganese oxides and particularly where copper stain is also present. Northwest of the San Juan mine are some rather low-grade, gold-bearing quartz veins associated with copper mineralization.

CLARK DISTRICT

The Clark district, of Graham County, is in the broad pass that separates the Santa Teresa and Pinaleno (Graham) mountains. Via the Pima-Klondyke road, it is 19 miles from Cork, a siding on the Southern Pacific Railway.

In this region, gold-bearing quartz veins have been prospected to a small extent since 1900 or earlier. Their production to the end of 1933 amounts to a few hundred tons of ore.

The prevailing rock is coarse-grained granite which floors a rolling pediment. In places, this granite is cut by bluish gray, chloritized dikes which weather yellowish brown.

The principal known vein of this vicinity occurs within a fault fissure that follows one of these dikes. This dike outcrops, with a width of one to 3 feet, for an intermittent length of more than 1,500 feet. The vein strikes N. 73° E., dips 60° to 80° NW., and is from one to 7 inches wide. It consists of coarse-grained grayish-white quartz with abundant cavaties which contain hematite and limonite.

During 1933, J. and G. Richey reopened an old 110-foot inclined shaft on the Chance claims and drove about 100 feet of drifts on the 50-foot level along the vein. In December, a car of sorted ore was being shipped.

RATTLESNAKE DISTRICT

The Rattlesnake district, of southwestern Graham County, is in the vicinity of upper Rattlesnake and Kilberg creeks, in the southeastern portion of the Galiuro Mountains. This area has yielded a small production of gold from the Powers, Gold Mountain, and Knothe properties. In 1932, according to the U. S. Mineral Resources, the district produced 71 tons of ore that yielded $597 worth of gold and 7 ounces of silver. Its 1933 production is reported to have been worth about $750.

This portion of the Galiuro Mountains is made up of a thick series of volcanic flows and tuffs, invaded by dikes and minor intrusive masses and rather complexly faulted. The region as a whole has been deeply dissected into blocky, steep-sided ridges, separated by canyons that generally follow fault zones. Rattlesnake Canyon, a northward-trending tributary of Arivaipa Creek, carries a small flow of water except during dry seasons and supports a sparse growth of pine timber. The district, is accessible from the Willcox-Arivaipa road by 20 miles of wagon trail that branches southward at Haby's ranch, 60 miles from Willcox and 47 miles from Safford. This trail is impassable for ordinary automobiles.

POWERS MINE

The Powers mine, held by the Consolidated Galiuro Gold Mines, Inc., is at the head of Kilberg Canyon, 20 miles by wagon trail south of Haby's ranch.

This property is reported to have been located about 1908. Prior to 1918, the Powers family and others worked the mine on a small scale and milled some ore with arrastres. In 1917, a 2-stamp mill was erected in Rattlesnake Canyon, 2 miles north of the mine, but never was operated. Small test lots of ore were shipped in 1932 and 1933. In 1933, the present company installed a small Ellis mill 5 miles north of the mine and, according to local reports, treated 100 tons of ore that yielded $7.50 worth of gold per ton.

Here, the prevailing rock is rhyolite, intruded by dikes of andesite-porphyry. The deposit occurs within a brecciated zone that strikes about N. 20° W., dips 55° W., and is about 100 feet wide. Its richest portion, which is from 3 to 15 feet wide, occurs near the hanging wall. The ore consists of brecciated, silicified rhyolite with small, scattered bunches of limonite and chrysocolla. The gold occurs as fairly coarse, ragged grains, mainly associated with the limonite and chrysocolla. The breccia shows considerable sericitization. Below the oxidized zone, the gold is probably contained in sulphides.

Workings on the Powers property include an adit that extends for 325 feet eastward, with about 100 feet of drifts in the ore-bearing portion of the brecciated zone, and an 80-foot winze with short drifts near the hanging wall.

KNOTHE MINE

The Knothe mine, held by E. Knothe, is one mile south of the Powers mine. It has yielded a few small shipments of sorted, high-grade gold ore. The prevailing rock in this vicinity is rhyolite, intruded by dikes of andesite-porphyry. The deposit occurs within a silicified, brecciated zone, up to about 3 feet wide, that strikes southeastward and dips steeply northeastward. Its richest portion is generally less than a foot wide and occurs near the hanging wall. The ore consists of silicified, brecciated rhyolite, cemented and partly replaced by sugary, grayish-white quartz and traversed by irregular veinlets of finely crystalline white quartz. It locally contains disseminations of fine-grained pyrite which probably carries most of the gold.

Workings on the Knothe property include several open cuts and a 40-foot shaft with short drifts.

GOLD MOUNTAIN PROPERTY

The Gold Mountain workings, in Rattlesnake Canyon, about 2 miles north of the Powers mine, were opened during the early years of the present century. They yielded only a small amount of gold ore and have been idle for many years.

Here, the prevailing rock is rhyolite, intruded by a few dikes of chocolate-colored andesite-porphyry. The gold-bearing ledge[264] dips about 72° W., in conformity with the adjacent rhyolite. It stands out sharply and has been cut by a branch of the canyon. As exposed, the ledge is separated into two parts each 20 or more feet wide by a thin body of rhyolite. It has been opened by short tunnels in the foot wall portion.

This gold-bearing ledge consists of granular quartz, traversed by scattered, irregular, thin veinlets of crystalline quartz. These veinlets are mineralized with gold-bearing pyrite. In the shallow workings, this pyrite has been oxidized to hematite and fine-grained, somewhat flaky, gold. The deposit is of rather low grade.

---

[264] Description abstracted from Blake, Wm. P., The geology of the Galiuro Mountains, Arizona, and of the gold-bearing ledge known as Gold Mountain: Eng. and Min. Jour., vol. 73, pp. 546-47, 1902.

PART II

THE OPERATIONS AT SMALL GOLD MINES

By John B. Cunningham,

Department of Mining and Metallurgy, University of Arizona.

INTRODUCTION

The price of gold in the United States has been normally about $20 per ounce for a hundred years. For the past thirty years, the prices of commodities have increased, at first slowly and then more rapidly during the World War and the after war boom. Especially during the last twenty years, with the large increase in the cost of things that the gold miner uses, he has been having a hard time—so difficult, in fact, that the old-time prospector was nearly eliminated. With the fixing of the price of gold, in January of 1934, at $35 per ounce, it has again become attractive for the prospector to seek the "hills" and look for gold.

In the interval of twenty or thirty years, many of the old prospectors have passed out of the picture and a new generation of men are again out looking for gold. Being inexperienced in such matters, the problem of developing their prospects is a troublesome one. Much of the information of the older men is not now available to the younger ones. With this fact in mind, the attempt is here made to collect from various sources, including some of the old-timers, information about the development of small mines, and to record these suggestions for the benefit of the "tenderfoot" now in the field.

Many gold mines, especially near the surface, are found in narrow veins. Most narrow veins, while they may be extensive, are irregular in form. The first principle to be followed in developing narrow gold veins, at least until the ore body has been proven to be of considerable extent and regularity, is to follow the ore shoot either by tunnels (horizontal openings) or shafts (vertical or inclined openings). A great deal of money has been wasted by the inexperienced in running crosscuts and in sinking shafts in barren ground with the idea of prospecting veins at depth. Much more valuable information can be obtained, more quickly, by following the ore.

When the topography is favorable, the cheapest method of following an ore shoot is by means of a tunnel. Drilling and blasting are easier, drainage is readily handled, and the removal of the broken rock is much cheaper. If the vein is narrower than the working width of the tunnel—3½ or 4 feet—it is an easy matter to mine separately the vein material and the country rock. A disadvantage that results from opening a mine by means of tunnels is the necessity of later sinking a shaft from the surface on the vein or sinking a winze in the mine, if it develops that the ore continues in depth.

Tunnels should be run with an even grade, that is not too steep, so as to make tramming as easy as possible. A grade of one inch to 1½ inches in 12 feet is a satisfactory grade for both tramming and drainage. A 12-foot board, 1 inch by 4 inches, with a block one inch to 1½ inches thick nailed to one end, if used with a carpenter's level, is useful in indicating the desired grade.

The position of the tunnel in relation to the vein will depend somewhat on the dip of the vein. As a general rule, the tunnel will be located so that the back of the tunnel will be in strong ground so as to lessen the amount of timber to be used. Quoting the United States Bureau of Mines information circular, "Mining and Milling Practices at Small Gold Mines," by E. D. Gardner, in regard to the location of tunnels with relation to veins: "If the vein is quite steep, the drift may be run equally well with the vein on either side or in the middle. If the vein dips 60° or less, it is more convenient to have the drift in such a position that the stope floor will intersect the side of the drift high enough to install chutes for loading a car or bucket. In flat veins, the drift is run as far under the vein as possible without losing it entirely." Figure 13 from the above Bureau of Mines circular shows the various positions of the drifts with relation to veins.

DRILLING

Mine openings, in the early stages of development, are usually made by drilling holes by hand. This saves the installation of expensive equipment such as compressors and machine drills, which often is not justified in small mines and prospects. The hand method of drilling is performed by an operator who uses four or five lengths of drill steel, usually ⅞ inch diameter. The following table shows the details of such a set of drills.

TABLE I

Hand Drill	Length of Drill	Gauge of Bit	Depth of Hole by Drill
1	6-in.	1¾ ins.	2 ins.
2	14-in.	1⅜ ins.	10 ins.
3	22-in.	1¼ ins.	18 ins.
4	30-in.	1⅛ ins.	26 ins.
5	38-in.	1 in.	34 ins.

Reductions in the gauge of the successive bits are necessary to enable the following drill to enter the hole. The latter diminishes in diameter gradually as drilling proceeds, due to the wearing of the bit. The number one drill is known as a starter and is usually just long enough to be easily handled. Some operators begin the hole by using a moil—a sharp pointed piece of steel.

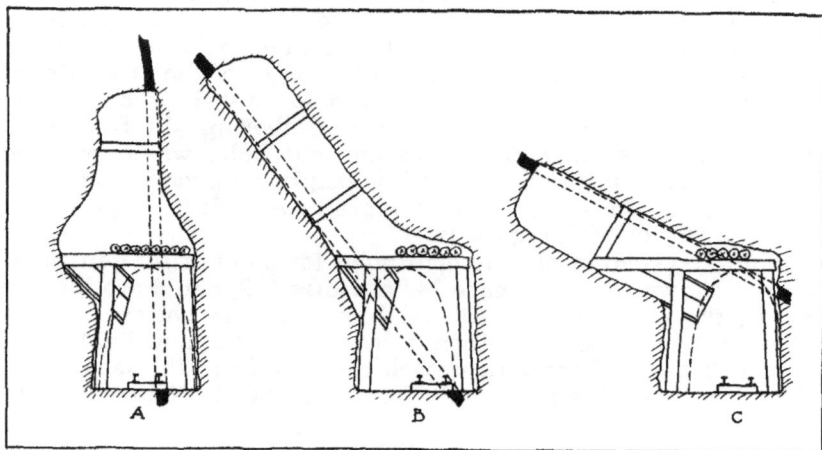

Figure 13.—Various positions of drifts with relation to veins and stopes.

The miner performs the operation of drilling by holding the drill in one hand, striking the head of the drill with the hammer held with the other hand, and rotating the drill a part turn between blows. Such an operation is known as single-handed or single jack drilling and the weight of hammer used is about 4 pounds. If one man holds and turns the drill and another strikes the blows, using a heavier hammer—6 to 8 pounds—, the operation is known as double-handed or double jack drilling, and such drilling is more economical in hard rock. Sometimes two strikers are used with a third holding the drill, in especially hard rock.

If possible, holes are pointed slightly downward so that a small amount of water may be kept in the hole. The small chips of rock are then kept in suspension and the drill strikes the solid rock. Periodically, the sludge that collects in the hole must be cleaned out with a small rod having a cup-shaped end.

Placing of drill holes: The first principle in the placing of drill holes, whether in open cuts or in small tunnels, is to take advantage of the natural conditions of the rock; that is, to place the hole in relation to planes of weakness and the contour of the surface so as to make blasting most effective. The hole should be placed across the bedding planes or other planes of weakness, as nearly as practicable. Experience soon teaches the miner, especially a hand-driller, how to place and point the holes.

In open cut mining, it is often possible so to plan the work that two faces are always exposed for blasting, as illustrated in "A," Figure 14.

In driving small tunnels, the center holes, called the cut holes, are blasted first, followed by the blasting of the holes around the outer edges, sometimes called the squaring-up holes. This procedure is illustrated in "D," Figure 14. In this case, there are three cut holes drilled towards a center to form a pyramid and seven squaring-up holes around the edges of the face. Small drifts driven in rather soft ground, should be well arched at the top so as to support the top without timbering. In Figure 14, "B," is shown such a drift with one cut hole, which may be drilled either pointing downward or pointing upward, and three outside holes. In harder ground, more holes will be required, as shown in "C" or "D," Figure 14.

Another system of placing drill holes for hand drilling in small drifts is shown in "E," Figure 14.[265] Holes 1, 2, and 3 are blasted first in the order numbered and blow out the bottom. Holes 4, 5, and 6 are fired next, then holes 7, 8, and 9, and finally, 10, 11, and 12, the last three turn back the rock from the face.

Hand drilled holes are usually from 30 inches to 36 inches deep.

DRILL STEEL

Steel used for hand drilling: This steel is usually solid, eight-sided steel, ⅞ inch in diameter, weighing 2 pounds per running foot, and containing about 0.7 per cent carbon. The drills are sharpened by hand after heating the end in a blacksmith's forge. Excessive heating harms the steel by decarbonizing it; that is, burning out the carbon which gives the steel its hardening quality. The best results are obtained by heating to a cherry red, turning in the forge to give uniform heating, and then forging on an anvil, turning after each half dozen blows. The blows should be light and glancing to draw the fibre of the steel and thus toughen the metal. A number of heatings and forgings will be required to shape the steel, probably five or six. During the later heatings, one is liable to overheat the end to a bright cherry red. If that is done, the end can be cut off and the new end again slowly heated to a low cherry red and forged to the final shape. Some operators finally grind the hot end to the desired edge with a file or grinding wheel. The angle of the edge varies from 65° to 85°. The sharper edged bit or cutting end cuts the rock faster but dulls quicker. Probably a 75° edge will be the best for average conditions. Experience will tell the operator

---

[265] Sketch E is from the Blaster's Hand Book, published by The DuPont Company, Explosive Department, Wilmington, Delaware, which book may be obtained free by writing to the Company.

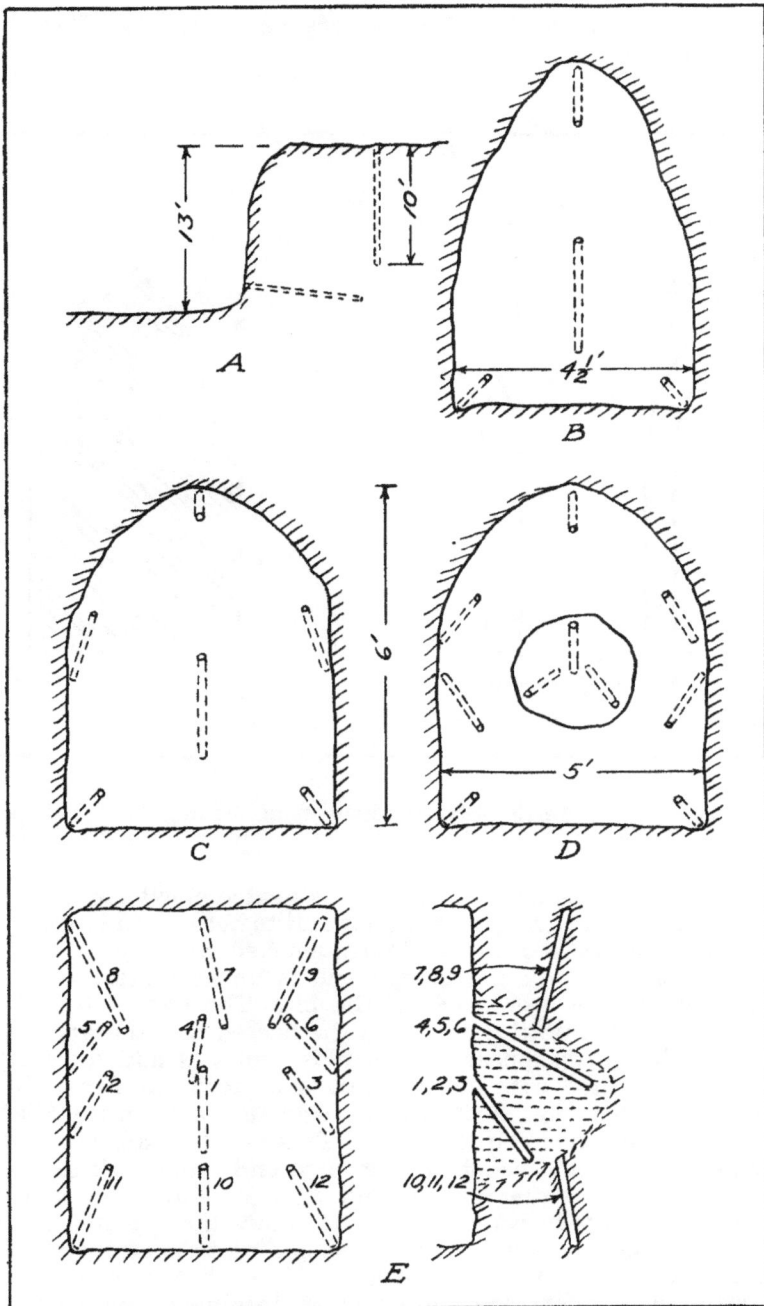

Figure 14.—The placing of drill holes for blasting

the kind of edge to make for the rock he is drilling. Figure 15 shows the shape of the starter bit.

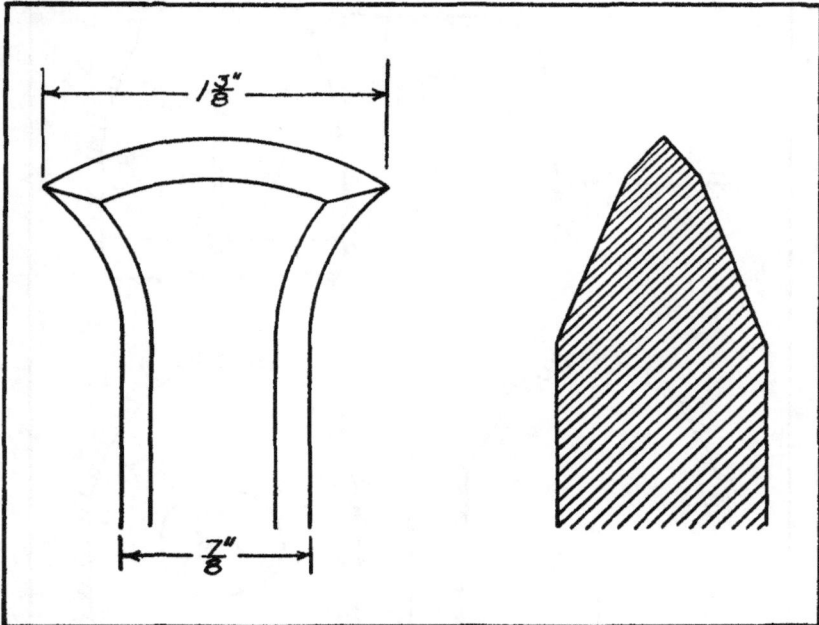

Figure 15.—Starter bit for hand drilling.

To harden and temper the steel, it is heated slowly so as to obtain an even heat throughout the bit and to between a low cherry red and a cherry red. It is then quenched by dipping in cold water so that half of the heated end is under water, which is about one inch or 1¼ inches of the bit. The end of the bit is held under water until the color of the heated part out above the water begins to disappear. Then it is removed and the forged surface is polished with emery cloth or a piece of emery wheel. During this operation, heat is being transferred from the heated part to the quenched end of the bit. This heat transfer is allowed to continue until the polished, quenched end comes to a shade of color between a straw yellow and brown yellow. It is then quenched quickly in water. Table II shows the various forging and tempering colors and their corresponding temperatures. The two temperatures marked with a star in each case show about the range of temperature and color for forging in one case and for tempering in the other.

TABLE II

Forging Heats	Tempering Heats
Low red heat — 975°F.	Very pale yellow — 430°F.
Dull red heat — 1290°	*Straw yellow — 460°
*Low cherry red — 1470°	(proper heat for drill steel)
(proper heat for drill steel)	*Brown yellow — 500°
*Cherry red — 1650°	Light purple — 530°
Clear cherry red — 1830°	Dark purple — 550°
Clear orange — 2190°	Pale blue — 610°
White heat — 2370°	Blue tinged with green — 630°
Dazzling white heat — 2730° to	
2910°	

EXPLOSIVES

Commercial explosives are solids or liquids that can be instantaneously converted by friction, heat, shock, or spark into very large volumes of gas. This increased volume of gas exerts pressure on the confining material. The pressure acts equally in all directions, but the gas tends to escape along the path of least resistance, or the easiest way out. The loading or placing of the explosives in the hole and the tamping or filling and packing, of the outer part of the hole not filled by the explosive, with clay or fine sand, must, therefore, be done very carefully in order to confine the gas and make it do the desired work on the rock to be blasted.

Various kinds of explosives are in use, depending on the kind and the place of the work to be done. For underground work, various grades of dynamite are in common use. They include straight, extra, ammonia, and gelatin dynamite. Dynamite consists chiefly of a mixture of nitroglycerin and some absorbent, such as wood pulp or saw dust, and an oxidizing substance (something that readily gives off oxygen on heating) such as sodium or ammonium nitrate. The gelatin dynamites contain, beside the above mentioned ingredients, a small amount of gun cotton dissolved in the nitroglycerin, which gives to this nitroglycerin a jelly-like character.

Various strengths of dynamite are sold, such as 20 per cent, 40 per cent, and 60 per cent, which figures indicate, in the case of straight dynamite, the percentage of nitroglycerin which they contain. Other dynamites that are marked with a percentage, such as 40 per cent, are made of such a mixture and content as will, when exploded develop the same strength as a 40 per cent straight dynamite. The percentage markings do not indicate the relative energy developed since a 40 per cent is not twice as strong as a 20 per cent dynamite, but only about 40 per cent stronger.

High explosives vary greatly in their power to resist water. In dry ground, this fact is of no importance, but, if much water is encountered, a water-resisting explosive is required. The gelatin dynamites are best under such conditions.

Fumes that are objectionable and sometimes dangerous are given off by dynamites or other high explosives when fired. In underground work, an explosive should be used that develops relatively little of these bad or dangerous fumes. The gelatin dynamites give off the least quantity of poisonous fumes.

The amount of bad fumes that are made by the explosion depends also upon several factors under the control of the miner. Burning dynamite makes the worst and the most dangerous fumes and incompletely detonated or fired dynamite is only slightly less undesirable. Complete detonation is obtained by confining the charge, tamping or packing it well with a wooden stick so as to remove air spaces, and using a sufficiently strong detonator. Sufficient material should be tamped down upon it so that there will be no chance for any part of the charge to blow out. Approximately one-half or a little more of the hole should be filled with the explosive leaving the other half to be filled with clay or sand, well packed in place. If the charge blows out or does not break the rock, the natural tendency is to use more explosive next time. What should be done is to confine the explosive more strongly, or use stronger explosives.

Old, deteriorated dynamite in some instances will fail to explode, causing misfires, and, at other times, will burn instead of detonating, giving off poisonous fumes. This is one reason for the State law that dynamite of a certain age must not be used and must be destroyed in the open.

Most high explosives made at present will not freeze under ordinary exposure to temperatures normally met with in this country and are called low freezing explosives. Blasting gelatin is not a low freezing explosive and, if it becomes frozen, which it may, it must be carefully thawed to give good results. Dynamite, when heated to 200° F., will explode as the result of a sharp shock or blow. When heated to 350° or 400°, it will explode from heat alone.

Nitroglycerin is readily taken into the body through the pores of the skin. It makes the heart beat faster and usually causes a headache.

Dynamite is supplied in the form of cartridges, usually called sticks, that are commonly of the following sizes: 1¼ inches diameter by 8 inches long, 1⅛ inches by 8 inches, 1 inch by 8 inches, and ⅞ inch by 8 inches. They are wrapped in oiled paper and supplied in boxes that contain 50 pounds net weight of explosives. Sometimes boxes that contain 25 pounds are sold. A 50 pound box of ⅞ inch, 40 per cent gelatin dynamite, will contain about one hundred and eighty cartridges.

Blasting Accessories: Accessories required when blasting include blasting caps, for detonating or firing the dynamite, fuse, a cap crimper, a tamping rod, and stemming material, besides the dynamite required.

Blasting caps are small copper cylinders, closed at one end, and, in the closed end, is a charge of a very sensitive explosive that is exploded by the spit or spark from the safety fuse placed in the empty end of the cylinder. Caps are commonly obtainable in two sizes; No. 6, which is 1.5 inches long by 0.234 inch in diameter and contains 1.0 gram of explosive; and No. 8 which is 1.875 inches by 0.24 inch and contains 2.0 grams of explosive. They are sold in boxes that each contain one hundred caps. A box of No. 6 caps costs about $2.00. The explosive in blasting caps absorbs moisture readily and decreases in sensitiveness when it becomes damp. Blasting caps should, therefore, always be kept in a dry place. The cover should be left on the box and the felt pad kept under it at all times, except when removing a cap from the box.

Safety fuse is made of a fine train of powder that is tightly wrapped in and waterproofed by outer and inner coverings of tape or twine. The standard fuse has a burning speed of either thirty or forty seconds per foot, depending on the brand. Fuse should be kept dry. It should not be bent sharply, crushed, or pounded. Oil and grease have a solvent action on the wrappings and may penetrate to the powder and delay or stop the burning. A cap crimper is a special tool for fastening the blasting cap to the safety fuse and for punching a hole in the cartridge of dynamite for inserting the detonator. The blasting cap must be securely fastened to the safety fuse to prevent the fuse from separating from the cap explosive while placing the primer in the hole and while the hole is being tamped. The primer is the cartridge of dynamite that contains the cap. The fuse sets off the cap, which in turn explodes the primer cartridge and, thus, the whole charge in the hole. Successful crimping can be done only with an instrument especially designed for that purpose. It makes the groove of a proper depth that will hold the fuse in place and not choke the powder train. The crimp is made near the open end of the cap. Crimping too near the other end might explode the cap. Figure 16 shows three methods of attaching the cap and fuse to a cartridge to make the primer. These figures are from a bulletin published by the Apache Powder Company which may be obtained for the asking.[266]

A few don'ts to be observed in handling explosives: Don't forget the nature of explosives, but remember that, with proper care, they can be handled with comparative safety.

Don't smoke while handling explosives and don't handle explosives near an open fire.

Don't shoot into explosives with a rifle or pistol.

Don't carry blasting caps in the clothing.

Don't attempt to take blasting caps from the box by inserting a wire, nail or other sharp instrument.

---

[266] Apache explosives, properties and characteristics, by Apache Powder Company, Benson, Arizona.

Don't leave explosives in a wet or damp place. They should be kept in a suitable dry place, under lock and key, where children and irresponsible persons cannot get at them.

Don't store dynamite in such a way that the sticks are on end, as that position increases the danger of the nitroglycerin leaking from the cartridge.

Don't store or handle explosives near a residence.

Don't open dynamite boxes with a nail puller, or powder cans with a pick axe.

Don't store or transport caps and explosives together.

Don't fasten a blasting cap to the fuse with the teeth or flatten the cap with a knife; use a cap crimper. The ordinary cap contains enough fulminate of mercury (high explosive) to blow a man's head or hands to pieces.

Don't put the fuse through the cartridge. This practice is frequently responsible for the burning of the charge.

Don't use force to insert a primer into a bore hole.

Don't do tamping (pressing down the stick of dynamite into the hole) with iron bars or tools; use only a wooden tamping stick with no metal parts.

Don't handle fuse carelessly in cold weather, since, when it is cold, it is stiff and breaks easily.

Don't cut the fuse short to save time. Such economy is dangerous.

Don't explode a charge before everyone is well beyond the danger line and protected from flying rocks. Protect the supply of explosives also from flying pieces.

Don't drill, bore, or pick out a charge that has failed to explode. Drill and charge another bore hole at least 2 feet from the missed one.

Blasting: Small prospect drifts that are driven by hand have rather shallow holes in comparison with those driven by power drilling machines. One and one-half sticks of dynamite to two and one-half sticks of 40 per cent gelatin dynamite are used, the amount depending upon the hardness of the rock to be blasted. In any case, the hole should not be filled more than two-thirds with the dynamite, and usually only one-half. In the ordinary hand drilled round, 30 to 36 inches deep, two sticks of dynamite will be the proper charge. Prepare the primer as previously explained and place in the bottom of the hole with the cap end pointing outward. On this primer place the second stick or a half stick if one and one-half stick are used. Press the whole charge down tightly with the tamping stick. On top of the charge place a wad of dry paper, leaves, or a rag. Press down lightly to avoid disturbing the cap. Then press into the hole some clay or moist, fine sand, nearly sufficient to fill the hole. While placing the tamping material (clay or sand), hold the fuse in one hand to keep it straight in the hole.

The length of fuse to be used for each hole can be calculated from the rate of burning and the time required for the miners to get to a safe place. Allow plenty of time. Such fuse to be spit or lighted should be cut open at the end so as to expose sufficient fuse powder for lighting. This may be done by cutting the end square across and then splitting the end for about one-half inch.

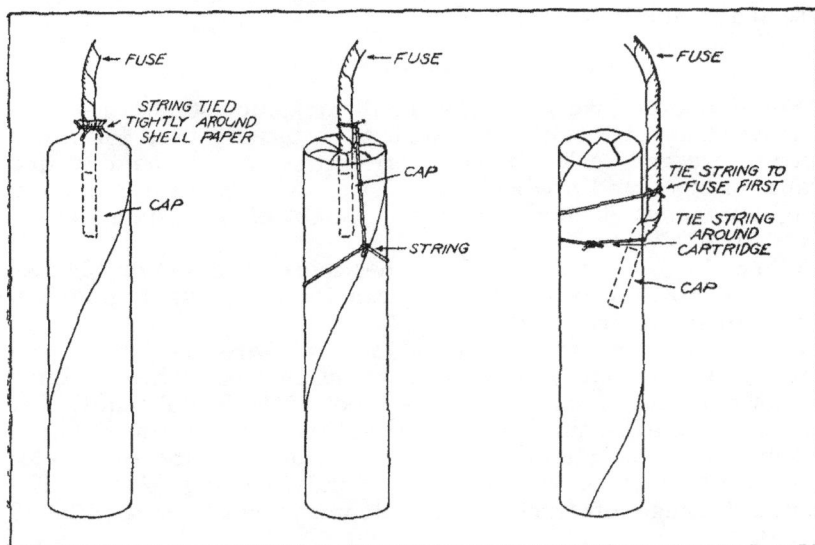

Figure 16.—Methods of attaching cap and fuse to the cartridge.

The fuses are lighted in the order in which the holes are to be exploded, the center or cut holes first, the top one next, and the bottom hole last.[267]

Misfires: This is a term applied to the failure of one or more of the shots to fire or explode. Whenever misfires occur, a very thorough investigation should be made. Never go back immediately, but wait several hours. The fuse may not be burning at its usual rate because of injury or some other cause. It may smoulder for hours, finally speed up, and fire the charge. Often a thorough investigation of the methods used in making up the primers, loading the holes, or in handling the fuse will reveal the cause of misfires. Cutting the fuse with a dull knife may smear enough of the water proofing material over the end of the fuse to interfere with the fuse spit firing the cap. Failure to place the end of the fuse in direct contact with the cap charge, when crimping the cap to the fuse, will leave a space where a gas pressure will build up, and that may prevent the firing of

[267] U. S. Bureau of Mines Bulletin No. 311, Drilling and blasting in metal mine drifts, and cross cuts — cost 40c.

the cap. Such a condition could be caused by not holding the fuse firmly in place when crimping the cap, or by not cutting the fuse off squarely. Lack of attention to the avoidance of an air-space between the fuse and cap charge has been the cause of many misfires. A damp fuse end may also give a weak spit. If no tamping has been used in a hole that has misfired, another stick of dynamite that contains a detonator (cap) can safely be inserted in the hole and exploded.

TRAMMING

For temporary work and for small, irregular stopes, a wheel-barrow of about 3 cubic feet capacity is usually used for trans-porting ore and rock. It has a pressed steel tray, handles and frame of steel, and a wheel guard of steel pipe. The maximum height is about 20 inches, it has a weight of 65 pounds, and it costs about $6.50.

When the face of a tunnel reaches more than 100 or 150 feet from the opening, a car with tracks for removing the broken rock will prove economical.

For development work, 8-pound steel rails are the lightest that can be obtained, but 12-pound rails are often used. These weights are for one yard of rail. A hundred feet of track will require 534 pounds of the 8-pound rails and, at about $3.50 per hundred weight, will cost $18.70. At the same rate, a hundred feet of 12-pound rail track requires 800 pounds, costing $28.00. The standard gauge of track for development work is usually 18 inches.

Ties can usually be cut from timber obtained in the neighbor-hood. They are made 2 or 3 inches by 4 inches and 30 inches long and are spaced about 18 inches apart. The upkeep of the track will be greatly reduced if, in laying it, the ties are set into the rock and ballasted so that every tie carries its proportional weight of track and car.

SHOVELING

Shoveling, being one of the ancient occupations, is not usually considered worthy of serious thought. Some attention is given to the kind of tool used, but very little to the process itself. Some years ago, a scientific study was made of the shoveling operation and it was demonstrated that a first class shoveler could do the most work with the least effort by using a shovel that would have a load of 21 pounds. It is evident, of course, that no shoveler can always take a load of exactly 21 pounds on his shovel, but, although his load may vary 3 or 4 pounds one way or the other, he will do his biggest day's work and be less tired when his average for the day is about 21 pounds per shovel load.

There are several kinds of shovels as regards quality, shapes and length of handle, and shape of blade. Handles are long or short and the end may be pointed or square. The blade may be round, half round, or square, and the shape of the blade may vary from flat to that of a scoop. The material to be handled, the place in which the shovel is to be used, the kind of work, etc., will govern the choice.

In working in small drifts without the use of a turn-plate (iron plate or wooden boards under the rock to be shoveled), experience seems to indicate that it is best to use a round-bladed shovel with a handle about 38 inches long. It is common practice to buy a long-handled shovel—these are 4 feet in length—and cut 10 inches from the end. If a turn-plate, sometimes called a sollar, is used, a flat, square-bladed shovel is best.

MINE SHAFTS

The smallest prospect shaft that is usually sunk is 4 feet by 5 feet. A shaft 4 feet by 6 feet or 4 feet by 7 feet is more convenient and is little more expensive. Shafts are seldom vertical as they should follow the vein as closely as possible so as to explore the ore body. Beside breaking the rock and ore, it is important to consider how best to hoist the broken material and how to provide timber to support the side walls in case they prove to be weak, which they usually are in veins. If timber is used, it will also furnish support for skids or guides for ore buckets and for ladders. If the walls are strong, some timber must be used for the support of these things.

To a depth of 8 feet, rock may be shovelled to the surface by hand, but, after that, a windlass is used. The economic limit for hoisting with a windlass is about 100 feet, but shafts have been sunk up to 300 feet deep by this method. Figure 17 shows a sketch of the common type of windlass used. A one-inch hemp rope is usually used with a light sheet iron bucket holding 150 to 200 pounds, having a capacity of about 2 cubic feet. One-inch hemp rope (3¾ feet per pound) costs about 35 cents per pound. Sometimes ¼-inch woven wire rope is used.

As greater depth is attained in prospect shafts, a head frame and a small hoist become necessary. Many small gasoline and oil hoists are now available, both new and second-hand. A new, 3-horse power, gasoline hoist can be bought for about $400 and a 6-horse power hoist for about $600.

Various kinds of head frames are in use, depending upon the depth of shaft, amount of ore to be hoisted, the inclination of the shaft, and the method of handling the ore or rock at the surface.

Figure 18 illustrates a tripod type of head frame for hoisting from depths of a few hundred feet. It can be constructed very quickly and cheaply and can be used with a small power hoist. When such a tripod is used, however, no special facilities for dumping the bucket at the surface are provided.

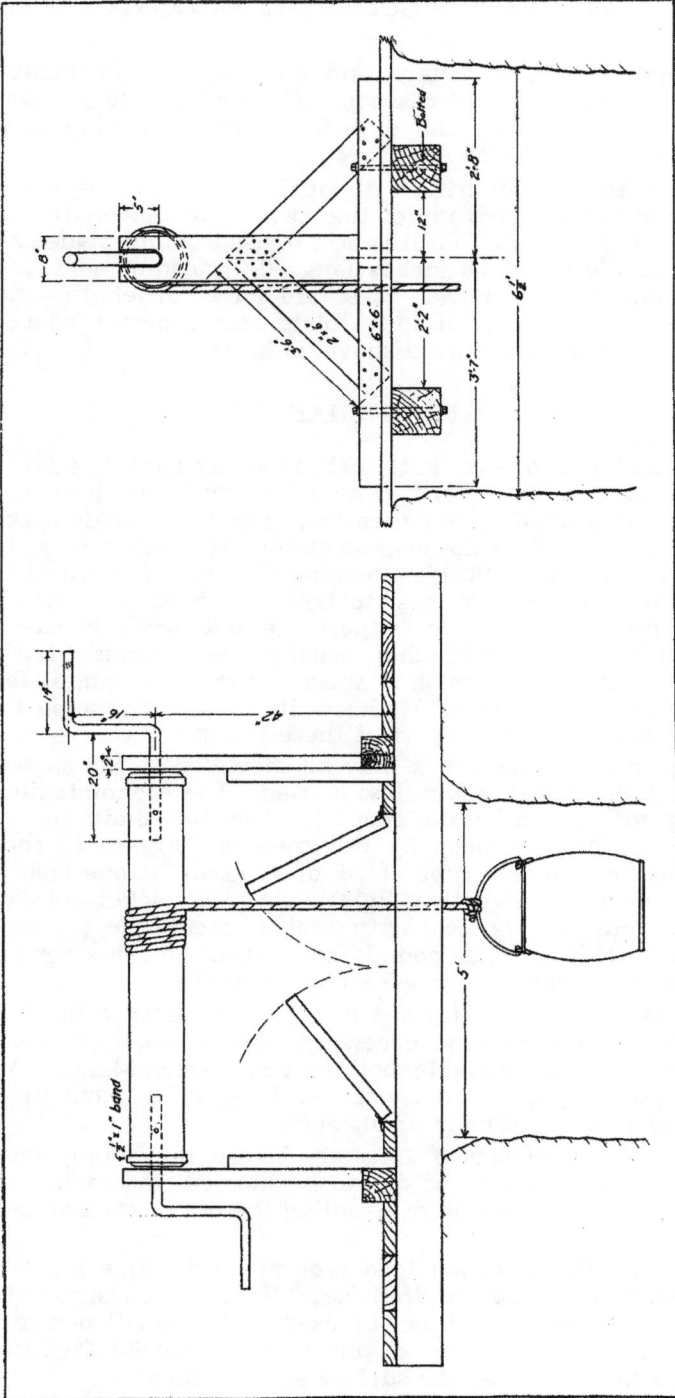

Figure 17.—Hand windlass.

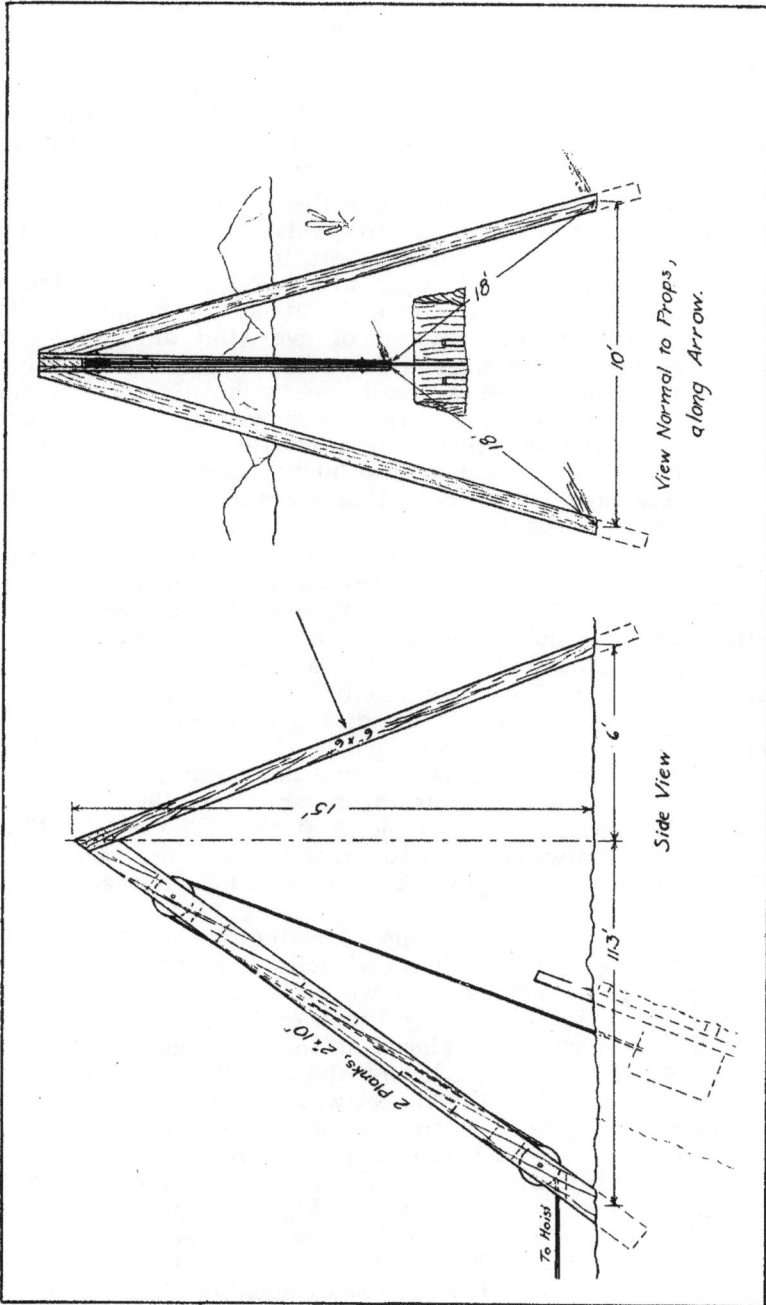

Figure 18.—Tripod type of headframe.

Another type of head frame in common use, with a self dumping device, is shown in Figure 19. This dumping device makes for speed in the handling of the bucket and lessens surface attendance. This head frame can be used for either vertical or inclined shafts. In the former case, skids (timbers on which the bucket slides) are easier to place in the shaft than are guides and it is unnecessary to use a cross head, often a source of danger.

The essential points that enter into the design of a head frame are the height of the bin or car to receive the hoisted ore and the distance of the hoisting drum from the shaft. The position of the sheave—the wheel or pulley at the top of the head frame to carry the winding rope—depends upon the height of the dumping device and the amount of overwind allowable. In prospecting shafts, the danger of overwinding is not great due to the fact that the hoist man can easily see the hoisting bucket, and the hoists usually used are slow, low-powered, geared hoists. Usually about 6 feet is sufficient to allow for overwind, and, allowing 6 feet for the length of the bucket, and the bale, and 2 feet of double cable at the bale, that gives about 12 feet as the distance from the dumping device to the sheave wheel.

Figure 19 shows side and end views of a standard head frame for a small prospect shaft. It consists of two nearly vertical posts, shown at "E," side view, in Figure 19, that are inclined slightly inward at the top as shown in the end view, and two back braces as shown at "D" in the same figure. To show the relative locations of the posts and braces to each other and to the hoisting drum and shaft, the dotted line "FB" is drawn to bisect—to divide into two equal parts—the angle at "F." This angle is formed by extending the two lines, one from the outside of the sheave to the hoist drum and the other from the sheave to the bucket, until they joint at "F." The point "B" of line "FB" should always fall between the foot of the back brace "D" and the foot of the post "E." It should be nearer to "D" than to "E."

Another thing to consider when locating the head frame in relation to the top of the shaft is the inclination of the shaft. As the bucket is being hoisted or lowered in an inclined shaft, it rides upon two skids (2 inch by 4 inch or 4 inch by 4 inch timbers) as runners which are placed on the foot wall of the shaft. In Figure 20, at "E," the bucket on the two skids is shown, the skids being placed so that the bucket will not touch the wall plate (the cross timbers placed on the foot wall to support the skids). In order that the rope shall remain parallel to the skids and the bucket properly ride the track at all times, the sheave for a shaft more inclined than the one shown in Figure 19 must be placed more to the left. The sheave will then be farther horizontally from the shaft, and, in other words, the foot of the nearby vertical posts will be farther from the shaft opening. The point "E" may be located by extending a line from the collar of the shaft

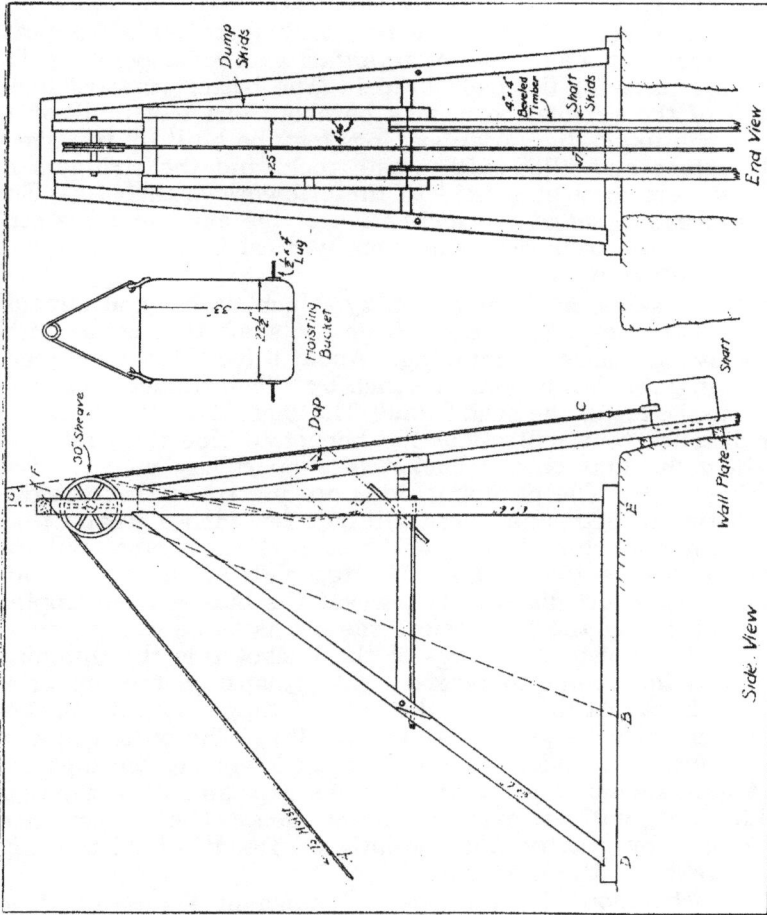

Figure 19.—Headframe.

to about the height of the sheave wheel, this line being at the same inclination as the shaft. From this sheave point, drop a plumb-line to the ground for the location of the foot of the vertical posts. The angle formed by a line extended from the shaft at its inclination, to above the sheave, with the vertical extension of the vertical post as shown at "G" will be the same as the difference between 90° and the dip of the shaft.

The nearly vertical posts are tied with girts (cross supports and tie rods) at about 10-foot intervals and corresponding distances are used for the back braces. The batter (inward inclination) of the vertical posts may be about one foot in 10 or 12 feet. For a foundation, ordinary logs may be built up to form a level surface and filled with waste rock, and the stringers of foot members may be spiked to the cribwork with bolts. The front braces are set at the inclination of the shaft and are connected with the back braces and the vertical posts by means of girts and tie rods.

The shaft skids, made of 4 inch by 4 inch timber and beveled on the inner, top edges, extend from the shaft to a point a few feet below the place of dumping. About 6 feet below the place of dumping, the dump skids of 4 inch by 6 inch timber begin and extend to the top of the head frame. The shaft skids are separated by a distance that will allow the bucket to ride the skids but not allow the lugs on the bucket to support any weight. See "E," Figure 20. The dumping skids are placed so that the distance between their inner faces will be a few inches greater than the largest diameter of the bucket so that the weight will be carried on the two side lugs. The top ends of the shaft skids are tapered a short distance to prevent the bucket from tipping while it is riding the lugs before the ore is to be dumped. As the bucket is hoisted, the lugs of the bucket ride the dumping skids until they come to recesses, called daps, on the top faces of the timbers. At that point, the hoist is stopped and the bucket allowed to swing by gravity, thus dumping. The bucket is now raised a few feet and then lowered. In lowering, the lugs engage trippers which are thrown over the daps and allow the lugs to slide down without catching in the daps. The tripper then swings back by gravity and the daps are freed to hold the lugs of the bucket on the next trip.

The construction of the tripper is shown in Figure 20. The side view "E," in this figure shows the bucket about to be lowered after discharging its load. The lugs of the bucket carry down the top of the tripper, which covers the daps allowing the bucket to be lowered. At the same time the bottom of the tripper swings up and is stopped by a bolt "B," so placed that the top of the tripper will cover the dap on the down swing. After the lugs pass the tripper, it swings back by gravity as one end is longer and heavier. The skids at the dap, which is about 2 inches deep, are lined with ⅜-inch by 2-inch iron, bolted flush with the timber. The tripper can be made of the same material and swung from a one-inch bolt.

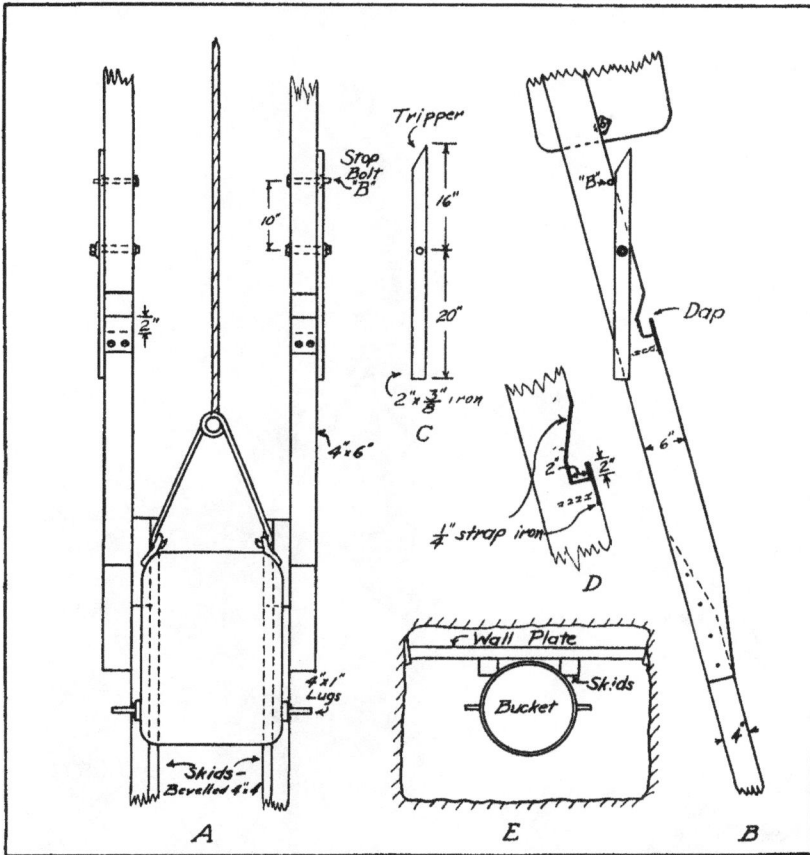

Figure 20.—Automatic dumping device for inclined shaft.

A bucket of the dimensions given in Figure 19 will have a capacity of about 8 cubic feet and will hold about 800 pounds of rock. Two lugs, 4 to 6 inches long and one-half to one inch in diameter, are fastened to each side about 8 inches from the bottom.

The headframe shown allows the bucket to dump directly into the car and is about 20 feet high and 16 feet between the feet of the vertical posts and those of the back braces. If an ore pocket is to be used and built into the headframe between uprights, the headframe may be built up to 40 or 50 feet in height.

Plate VI shows a headframe at an Arizona mine.

Plate VI.—Headframe at an Arizona mine.

The construction of an automatic device that is in common use in sinking vertical shafts and winzes is shown in Figure 21. Attached to the bottom of the bucket is an 18-inch, heavy chain to which is attached a 4-inch iron ball. As the bucket is hoisted above the dumping device, a dumping platform is swung from the vertical position as shown in dotted lines so as to lie at an angle across the shaft as shown in heavy lines. On the inclined platform rests a framework that supports a plate with a V-shaped notch. As the bucket is lowered, the ball catches in the notch and the bucket is tipped, emptying the rock down the chute into a bin or car. The bucket is then raised, the ball releases itself from the V-notch, and the whole device is raised to the original vertical position by means of a rope that is handled by the hoist operator. The bucket can then be lowered.

HOISTING ROPE

It is usual to use manila rope when hoisting with a windlass. The following table gives the relation between the size, feet and inches per pound, and the weight required to break the rope of some of the commonly used best manila rope. Using a safety

TABLE III

Dia. in inches	Length per pound	Breaking strength in pounds
¼ inch	48' — 0"	450
½ inch	12' — 0"	1800
¾ inch	6' — 0"	4000
1 inch	3' — 7"	7100

factor of five, no load greater than one-fifth of that given in the table should be used on new rope. For a one-inch rope, therefore, the maximum load should not be over 1400 pounds.

Wire rope for hoisting, with diameters of ¼ inch, ⅜ inch, and ½ inch, may be obtained through local dealers. It usually has six strands of nineteen wires each and is flexible as well as strong.

The following table[268] gives data concerning four sizes of wire

TABLE IV

Diameter inches	Wt. per foot lbs.	Approx. breaking strength, tons	Minimum dia. of sheave, feet
¼	0.10	2.2	1.0
⅜	0.22	4.8	1.5
½	0.39	8.4	2.0
⅝	0.62	12.5	2.5

---

[268] Peele's Mining Engineers Hand Book. John Wiley & Sons, New York.

rope. Again a safety factor of not less than five (no load greater than one-fifth of the breaking strength) should be placed on new rope. For old rope, the safety factor used must be greater; say six or eight.

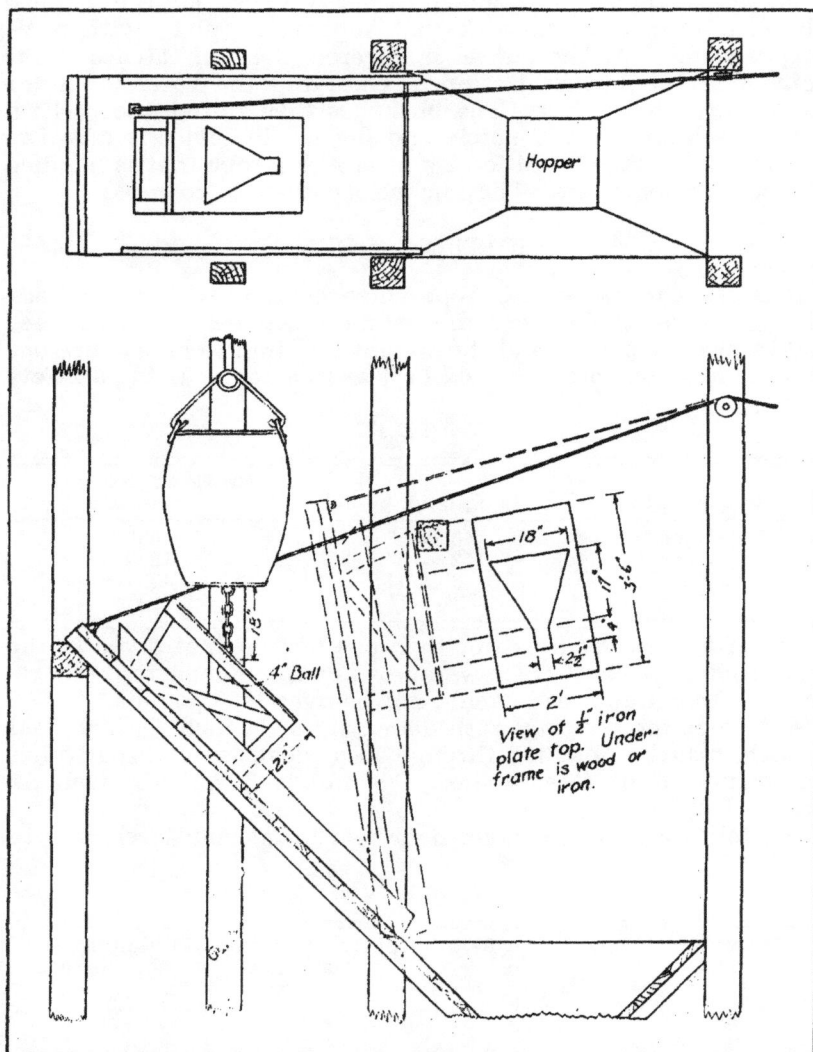

Figure 21.—Automatic dumping device for vertical shaft.

TIMBERING

Drift Sets: In driving small drifts and crosscuts, the ground is usually strong enough, especially if the roof is arched, so that it will not be necessary to use timber; but often zones of fracture and weak ground are encountered, and it is then necessary to use timber to prevent cave-ins.

As no two mines are exactly alike, and, as different kinds of rock require different methods of support, the miner must always use his judgment in selecting a method to hold up the rock. Figure 22 shows some examples of rock supports in drifts. The simplest form of roof support is a single timber laid horizontally and supported at the ends by hitches or recesses in the side walls. It is wedged at one end to hold it firmly. Sketch "A" shows such a support. If either or both side walls are not strong enough to support the ends of such a single timber, the ends may be supported by posts set vertically or slightly inclined inward at the top. A simple method to keep the tops of the posts in place is to nail a 2-inch plank to the bottom of the top member or cap. This saves forming notches to hold the posts. Sketch "B" shows the arrangement of this 3-piece set. When the bottom of the drift is not strong enough to support the weight of the posts a horizontal cross timber or sill is placed for them to rest upon. It is best to notch the ends of this sill to keep the bottoms of the posts in place. Sketch "C" shows this 4-piece set. These sets are placed about 5 feet apart. Between sets, above the caps, are placed small poles or planks, called lagging, to hold the loose rock. In swelling or loose ground, lagging may be put in back of the posts at the sides.

In order that these drift sets may not be forced out of place by the pressure of the rock (moved toward each other in either direction), small timbers are placed horizontally at both top and bottom corners from set to set. They are so placed as to rest against both cap and post at the top and both sill and post at the bottom corner. These horizontal members are sometimes called sprags. If the drift sets are not properly spragged, a slight movement of the ground may cause a set to collapse, weaken others near it, and lead to a cave-in of the drift.

If the drift is following an inclined vein from which it is expected, later, to mine the ore and form a stope, it is best, in fairly strong ore, when using a one-piece timber or stull for support, always to set this member at right angles to the vein. If one of the walls of the vein is weak it is often desirable to place a post to support that end of the stull. The other end can then be supported in a hitch in the strong wall. Figure 23 illustrates the placing of the one- and 2-piece sets under such conditions.

Miners in the western states usually prefer those timbers commonly known as spruce and Oregon pine, the latter being sometime known as Douglas spruce, Douglas fir, or Yellow fir. In

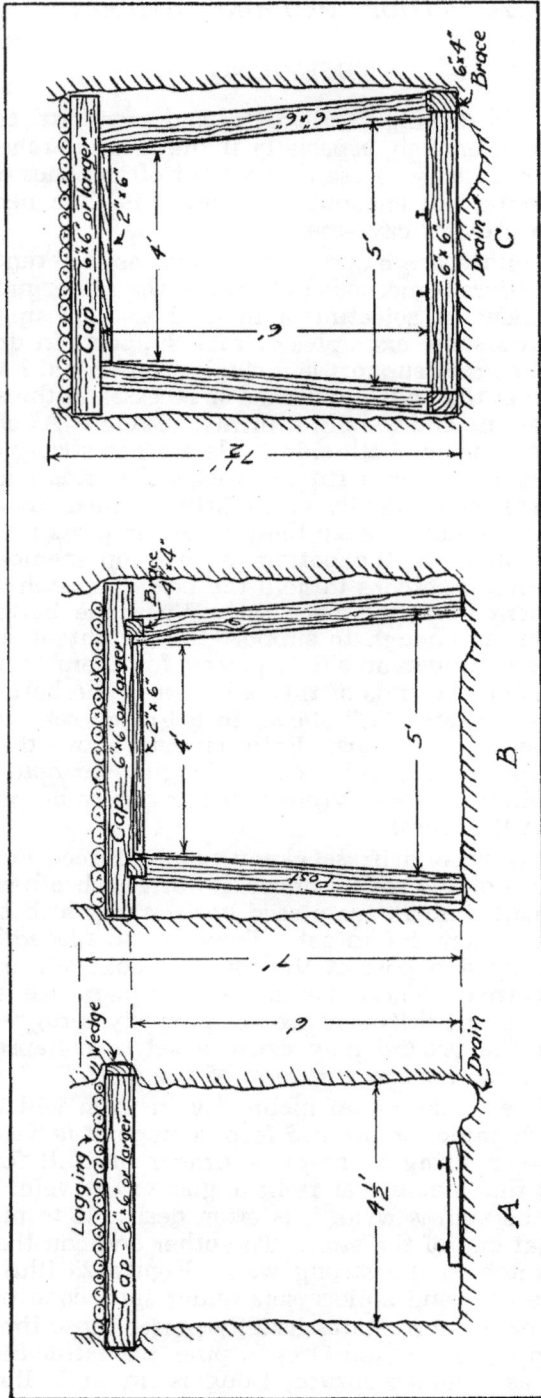

Figure 22.—Timber supports for small drifts.

addition to these two kinds of timber, yellow pine, sugar pine, various varieties of fir, and also oak are used.

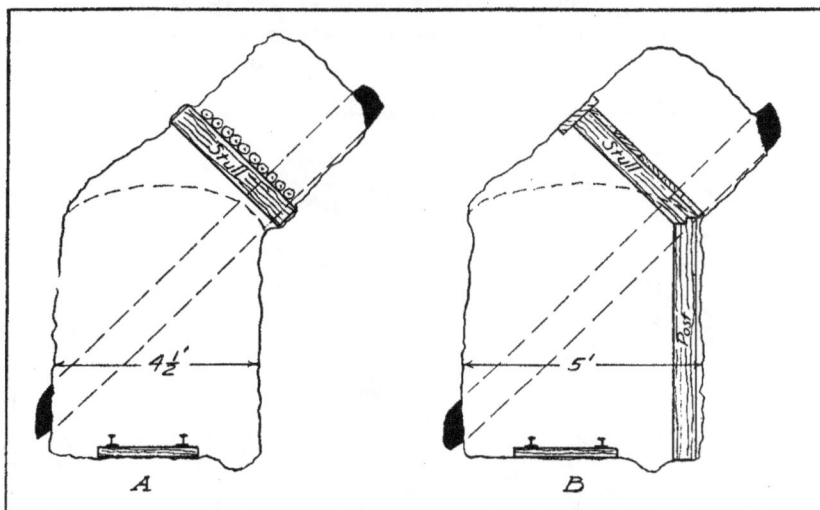

Figure 23.—One- and two-piece timber supports.

In a desert country, yucca, juniper, cottonwood, scrub pine, old railroad ties, and, in fact, almost any kind of wood available has been used by prospectors in timbering their workings.

Shaft timbering: In many prospect shafts, the rock is sufficiently strong so that it is not necessary to use timber. When the ground begins to loosen, stulls (timber placed perpendicular to the direction of rock movement) may be placed. In inclined shafts, sunk in strong ground, the only timbering necessary will be that to support the skids on which the hoisting bucket rides. Wall plates of 6 inch by 6 inch timbers are wedged into place on the foot wall and may be placed 6 to 8 feet apart. It is advisable to place the ends of some of these wall plates in hitches or recesses in the side walls every 30 or 40 feet so as to carry the down thrust of the members. Figure 24 shows a plan, and side view of the members in an inclined shaft, 7 feet wide and 4 feet across.

Figure 25 shows the details of the members of a shaft set that is used in prospect shafts, either vertical or inclined, when it is necessary to support the wall rock of the shaft opening. The set shown is for a small two-compartment shaft having a hoisting compartment that is 4 feet by 4 feet, inside dimensions, and a smaller compartment, 3 feet by 4 feet, for a ladder and any pipes that may be required, such as those used for water, air, and ventilation.

For erecting these sets, four iron hangers are used to support the lower wall plates while placing the other members. The whole set is then wedged to the side walls. These hangers are usually made in two pieces, threaded at one end and having a hook at the other. This makes it easier to put them in place and to remove them, than if they were of one piece and threaded at both ends. Large, thick washers are always used, with heavy nuts.

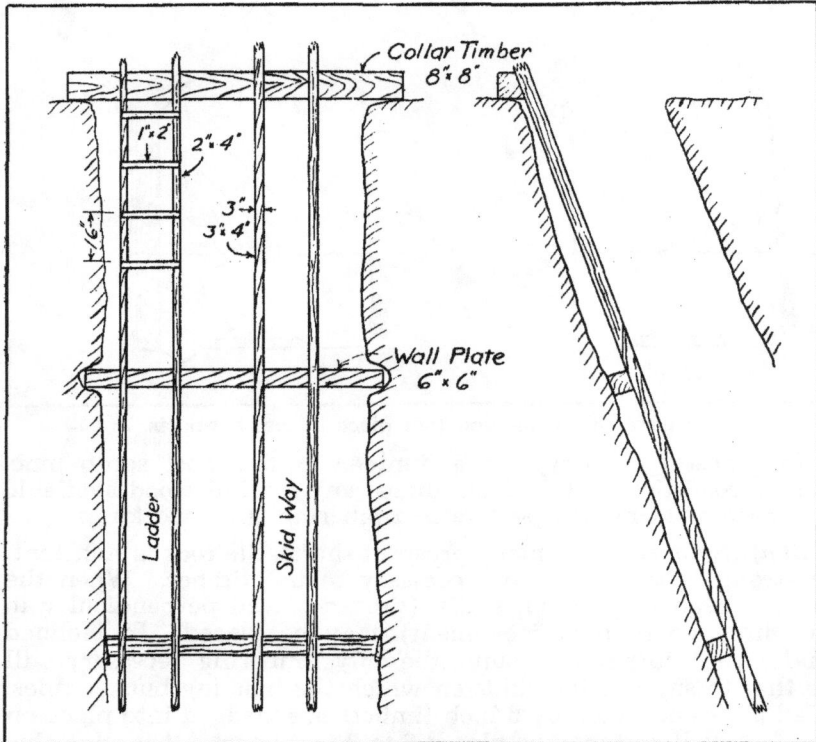

Figure 24.—Supports for shaft skids.

Wedging alone is not sufficient to support the shaft sets. Every 30 or 40 feet, in a prospect shaft, two long, heavy timbers, say 10 inch by 10 inch, must be placed across the shaft, under the ends of the wall plates, their ends extending into hitches or recesses in the wall rock. They are known as bearing timbers. The end plates are set as usual. The only difference in the set members that must then be made is in the corner posts which will be shorter by the thickness of the bearing member.

Sometimes the cut side of the shaft set is boarded up to protect the workmen and equipment from falling rock. These boards, called lagging, are usually made of 2 inch by 12 inch plank. They are held in place by cleats, 2 inch by 2 inch, nailed on the outside of the wall and end plates, before the plates are put in place, and also wedged.

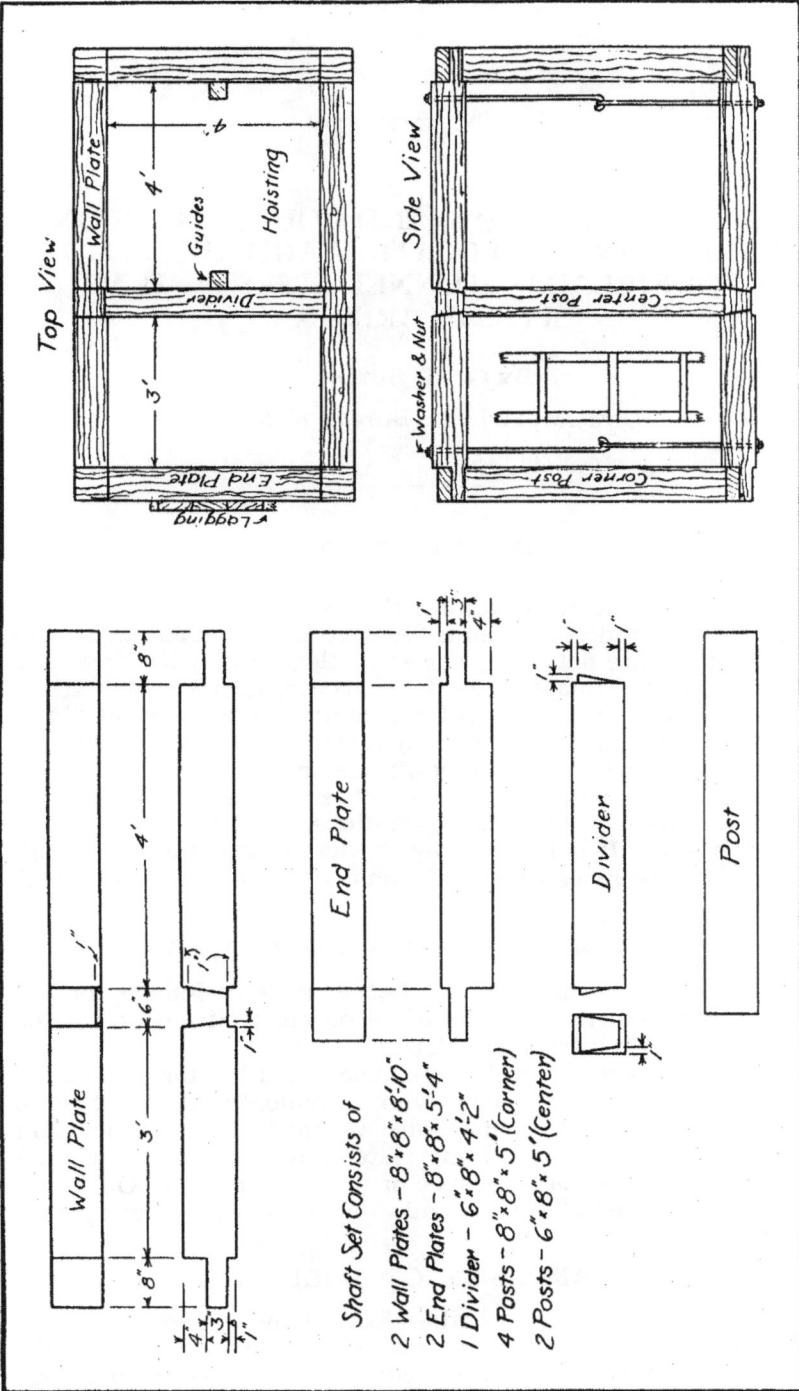

Top View

Wall Plate

Guides

Hoisting

4'

3'

End Plate

Divider

Lagging

Side View

Center Post

Washer & Nut

Corner Post

Wall Plate

8"

4'

1"

6'

1½"

3'

8"

5"

⅞"

¾"

1"

End Plate

1"

3"

4"

Divider

1"

1"

Post

Shaft Set Consists of

2 Wall Plates – 8"×8"×8'-10"
2 End Plates – 8"×8"×5'-4"
1 Divider – 6"×8"×4'-2"
4 Posts – 8"×8"×5'(Corner)
2 Posts – 6"×8"×5'(Center)

Figure 25.—Details of shaft timber sets.

PART III

LAWS, REGULATIONS, AND COURT DECISIONS BEARING ON THE LOCATION AND RETENTION OF LODE CLAIMS, TUNNEL SITES, AND MILL SITES IN ARIZONA

By G. M. BUTLER,

Director, Arizona Bureau of Mines

INTRODUCTION

The Federal and State statutes that cover the location and retention of lode and other claims required in connection with mining operations are not always entirely clear, and it has frequently been necessary to appeal to the courts for decisions on various questions relating thereto. Unfortunately, the decisions of different courts sometimes seem to be at variance with each other, and it is then impossible to state positively the meaning of the law. The statements that follow are either transcriptions of the statutes themselves or direct quotations, or they are based on court decisions. It is believed by the writer that they are trustworthy, but he recognizes that some of them are matters of opinion.

WHO MAY LOCATE LODE CLAIMS

A citizen of the United States (either male or female) or one who has declared his intention of becoming a citizen before the proper court may locate lode claims.

An Indian or a minor child may make valid locations as may a group or association of people who individually are entitled to locate claims. Locations may also be made by a corporation chartered under the laws of any state or territory of the United States, but no officer or employee of the General Land Office, including clerks, special agents, and mineral surveyors, can locate such claims.

SHAPE AND SIZE OF LODE CLAIMS

A lode "mining claim, whether located by one or more persons, may equal, but shall not exceed, fifteen hundred (1,500) feet in length along the vein or lode.............. No claim shall extend more than three hundred (300) feet on each side of the middle of the vein on the surface............... The end lines of each claim shall be parallel to each other."

A rectangular claim 1,500 by 600 feet contains 20.66 acres, and no single claim can have a larger area. It is not essential, however, that all claims be of that size, since claims may be any length less than 1,500 feet and any width less than 600 feet. (See Figure 26, c.) It is easiest and simplest to take up claims of rectangular form, but it is not necessary that they be rectangular. All the claims shown on Figure 26 are valid in Arizona.

THINGS TO BE DONE TO LOCATE A LODE CLAIM

1. Find a "vein," "lode," "ledge," or "crevice" containing "mineral in place." The term "mineral" means material of value that may be mined profitably if found of suitable grade and in sufficient quantity. If such material is a source of a metal or metals, it is commonly called ore. It is not necessary that ore in place that can be mined profitably be exposed at the point of discovery, but the conditions found there must be such as will lead an experienced miner to believe that good ore in such quantities as can be mined profitably might be found with further exploration. Many claims are illegally located because absolutely no ore or gangue material and no conditions that might develop into an ore body with further exploration are found where the discovery shaft or cut is sunk or excavated.

The discovery of "mineral in place" may be made by sinking a shaft or excavating a cut, but it is often the result of observing material that outcrops on the surface.

The finding of "float" (broken fragments of ore and gangue), even if present in such an amount that it may be collected profitably, does not constitute a discovery of "mineral in place," and a lode claim cannot be located on such material.

2. Erect at or contiguous (actually touching or closely adjoining) to the discovery a "conspicuous monument of stones, not less than 3 feet in height, or an upright post, securely fixed and projecting at least 4 feet above the ground."

3. Prepare a "location notice," in duplicate, and place one copy on the post or in the stone monument mentioned in the preceding paragraph. If placed on the post, it is customary to fasten a tight wooden box, open at one side, to the post, and to tack the location notice to the inside of that side of the box that is against the post, thus protecting the notice from the weather. If a stone monument is used, the location notice is usually put into a tin can that has a cover, and the can is laid with the rocks in the upper part of the monument. A location notice is said to have been "posted" if affixed to a post or placed inside a stone monument.

The "location notice must contain: the name of the claim located; the name of the locator or locators, the date of the location; the length and width of the claim in feet, and the distance

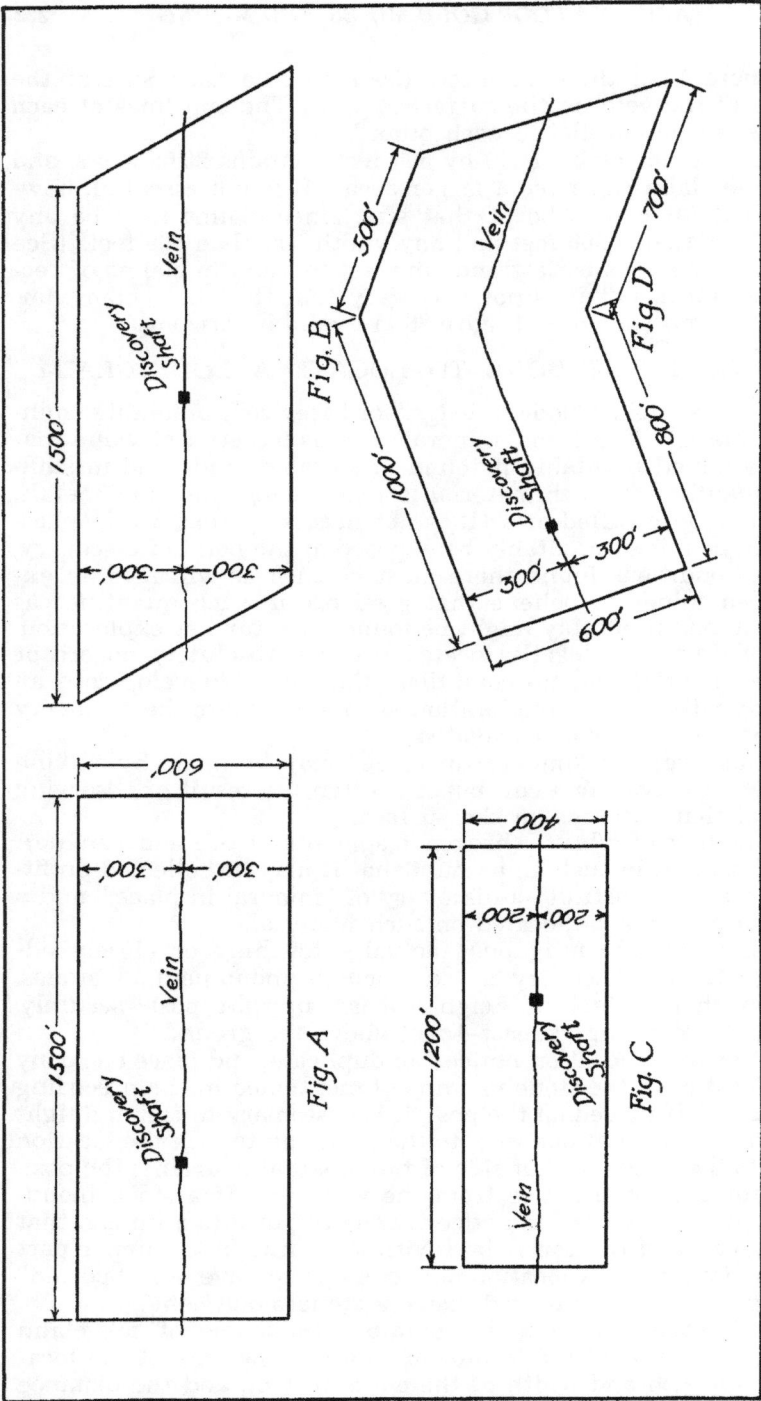

Figure 26.—Examples of lawful lode mining claims.

in feet from the point of discovery; the general course or courses of the claim; and the locality of the claim with reference to some natural object or permanent monument whereby the claim can be identified."

The Arizona statute just quoted makes it necessary to use longer and more detailed location notices than are required in many states.

Blank location notices may be obtained at good stationery stores, and, often, printing establishments, and these printed notices are commonly used. There follows, however, a sample location notice which complies with all the requirements of the Federal and state statutes. The words in italics must, of course, be changed to meet the existing conditions.

Prosperity Lode Claim

We, the undersigned, locators of this, the *Prosperity* mineral bearing lode which was discovered by *us* on *September 10, 1933,* claim 300 feet on each side of the center of the vein, *500* feet in an *easterly* direction, and *1,000* feet in a *northerly* direction from this discovery point from which the *junction of Spring and Mill creeks* bears *north easterly about a quarter of a mile and Mineral Monument No. 45* bears *south easterly about 1,000 feet.*

John Smith and Fred Jones.

The location notice should be posted as soon as possible after a discovery of "mineral in place" has been made and it may be posted prior to the actual discovery in order to protect seekers for mineral while they are searching for ore. The discoverer of the lode is not protected by the law until the location notice has been posted.

Any number of individuals may jointly locate a claim and any one of them can act as agent, without written authority of any or all the others, signing his or their names, as well as his own, to location notices for posting or for recording.

4. Within ninety days from the date of the discovery, have a copy of the location notice, attested before a notary, recorded in the office of the Recorder of the county in which the claim is located. It is not necessary that the location certificate that is recorded by the County Recorder be an exact duplicate of the location notice that was posted although it must convey all the information given by the posted location notice. It is common practice to use the brief form of location notice given for posting purposes and to use a certificate obtained from a stationer or printer for recording. The latter is considerably longer than the brief form of location notice given and the bearings and lengths of all the boundaries may be inserted therein.

5. The Arizona statutes provide that, within ninety days from the time of the location of a lode claim, the locator shall "sink a discovery shaft in the claim to a depth of at least 8 feet from

the lowest part of the rim of the shaft at the surface, and deeper, if necessary, until there is disclosed in said shaft mineral in place. Any open cut, adit, or tunnel made as part of the location of a lode mining claim, equal in amount of work to a shaft 8 feet deep and 4 feet wide by 6 feet long, shall be equivalent, as discovery work, to a shaft sunk from the surface."

6. Within ninety days from the date of location, "monument" the claim by setting "substantial posts projecting at least 4 feet above the surface of the ground" or by erecting "substantial stone monuments at least 3 feet high at each corner or angle of the claim and in the center of each end line," where the lode presumably passes out of the claim. The posts or stone monuments should be numbered consecutively, beginning with any one of them, and it is desirable to give the name of the claim on or in each of them.

Failure to do any of the previously mentioned six things constitutes, in effect, an abandonment of the claim, and the rights or claims of the locator are thereby lost.

CHANGING OR AMENDING LOCATIONS

A location notice may be amended at any time and the monuments may be changed to correspond with the amended notice *provided* that no change is made that will interfere with the rights of any one else.

LOCATION OF ABANDONED OR FORFEITED CLAIMS

The Arizona statutes provide that "the location of an abandoned or forfeited claim shall be made in the same manner as other locations except that the relocator may perform his location work by sinking the original shaft 8 feet deeper than it was originally, or, if the original location work consisted of a tunnel or open cut, he may perform his location work by extending said tunnel or open cut by removing therefrom 240 cubic feet of rock or vein material."

NUMBER OF CLAIMS THAT CAN BE LOCATED

There is no limit to the number of claims that may be located by an individual or group of persons, but all of the six things heretofore mentioned as essential must be done in connection with the location of *each* claim.

ANNUAL LABOR OR ASSESSMENT WORK

The Federal statutes provide that "on each mining claim.........., until a patent has been issued therefor, not less than $100 worth of labor shall be performed or improvements made during each year," and, upon failure to comply with this condition, "the claim or mine upon which such failure occurred shall be open to re-location in the same manner as if no location of the same had ever been made, provided that the original locators, their heirs, assigns, or legal representatives, have not resumed work upon the claim after such failure and before such location."

The work that is done in order to retain a claim under this provision of the law is usually called annual labor or assessment work, but it is also occasionally termed representation work.

The Assessment Work Year: The assessment work year during which the anual labor on claims must be performed begins and ends at noon on the first day of July.

Annual Labor not Required During the Assessment Work Year in which the Claim is Located: No annual labor is required on a claim until some time during the year following the first day of July next succeeding the date of location of the claim. As an illustration, assume that a claim was located on August 15th, 1933, and a discovery shaft was completed on November 1, 1933. No assessment work is required on that claim until some time between noon of July 1, 1934, and noon of July 1, 1935.

It is the *date of location* that is significant, however, and not the date when the discovery shaft is completed. If a claim were located on June 15, 1934, for instance, but the discovery shaft was not completed until July 14, 1934, assessment work would have to be done before noon of July 1, 1935, unless suspended by Congress.

Suspension of Assessment Work: During the three years that preceded July 1, 1934, Congress passed bills suspending assessment work on some mining claims. During 1931-1932, all such work was suspended and no one was required to do it; during 1932-1933 and 1933-1934, the privilege of refraining from doing such work was extended only to people who were exempt from the payment of a Federal income tax; and, during 1933-1934, a maximum of six lode claims or 120 acres of placer ground could be held by an individual, while a maximum of twelve lode claims or 240 acres of placer ground could be held by a partnership, association, or corporation without doing the annual labor.

Whether subsequent Congresses will pass similar legislation and what provisions will be included in such laws as may be passed cannot be foretold, but the Arizona Bureau of Mines is always ready to advise prospectors or miners about such matters.

Assessment Work Begun but not Completed: If a claim owner begins to do his assessment work at any time prior to noon of July 1 and uses due diligence in completing that work, even though it may not be finished for several days or weeks, he is considered to have complied with the law, but he must again start to do his assessment work before noon of the next succeeding July 1.

Title to Claims not Always Lost by Failure to do Assessment Work: Although failure to do assessment work on a mining claim within the time specified by law makes it possible for anyone to go on the ground and locate the claim as if no location of same had ever been made, the original locator does not lose *all* title to the claim unless someone else locates it. Even though he may have abandoned the claim for several years, he may do assessment work at any time, provided no one else has already started location proceedings on that claim, and the original locator's title to it is thereafter just as good as though he had continuously complied with all the provisions of the law that relate to the performance of annual labor.

Assessment Work on a Group of Lode Claims: "When a number of contiguous (lying side by side or end to end) claims are held in common (by one owner or the same group of owners), the aggregate expenditure that would be necessary to hold all of the claims may be made upon any one claim" or any fraction of the number of claims in a group, *provided* the work done or the improvements made manifestly benefit or facilitate the development of *all* of the claims in the group.

Cornering (touching corner to corner but not side to side or end to end) locations are held *not* to be contiguous.

Recording Affidavit of Performance of Annual Labor: Neither the Federal nor the Arizona statutes *require* the owner of a mining claim to file or record an affidavit that he has performed the assessment work, but the Arizona statutes provide that such an affidavit *may*, within three months after the time fixed for the performance of the annual labor or the making of improvements upon the mining claim (up to noon of 1st of October), be filed or recorded in the office of the Recorder of the county in which the claim is located, and, when so filed, it shall be *prima facie* evidence that the annual labor has been performed. The Arizona statutes provide, also, that this affidavit shall be in substance as follows:

"State of Arizona, County of........................ss:

...., being duly sworn, deposes and says that he is a citizen of the United States and more than twenty-one years of age, resides at.............................., in.............................. County, Arizona, is personally acquainted with the mining claim known as........................mining claim, situated in.................... mining district, Arizona, the location notice of which is recorded in the office of the County Recorder of said county, in book............ of records of mines, at page............ That between the............day of..........., A.D..........., and the............day of..........., A.D..........., at least............................dollars worth of work and improvements were done and performed upon said claim, not including the location work of said claim. Such work and improvements were made by and at the expense of............................, owner of said claim, for the purpose of complying with the laws of the United States pertaining to assessment or annual work, and (here name the miners or men who worked upon the claim in doing the work) were the men employed by said owner and who labored upon said claim, did said work and improvements, the same being as follows, to-wit:　(here describe the work done, and add signature and verification).”

The Arizona statutes further provide that:　“When two or more contiguous claims are owned by the same person, and constitute a group, and the annual work is done upon each of said claims, or upon one or more of the same for the benefit of all, all such claims may be included in a single affidavit.”

No Assessment Work Required After Application for Patent has been Made:　After an application for a patent has been filed, no more assessment work need be done no matter how long a time may elapse before patent is issued, providing the claimant and his attorney exercise due diligence in securing the patent.

Permissible Assessment Work on Lode Claims:　The Federal or State statutes do not specify what work will satisfy the requirements of the annual labor law.　Court decisions furnish the only guide in this matter, and they, naturally, do not cover all of the questions that might be propounded.　Some questions can be answered positively only by carrying the matter through the courts, and one man’s opinion is as good as another’s until the courts have spoken.

From court decisions, which other courts might reverse, we learn that the cost of doing certain things on or in connection with lode claims may be considered as probably fulfilling the annual labor provisions of the law, as follows:

1.　Digging, deepening, or extending a prospect shaft, tunnel, adit, open cut, drift, or cross-cut.

2.　Erecting structures or installing machinery needed for mining, but not for milling or smelting.

3.　Constructing any road-way, either on or off the claim, or a building if same is built *exclusively* in order to benefit that claim and is necessary in order to develop or to mine on it.

4.　Making diamond or churn drill tests.

5.　Employing a watchman to take care of or to protect mine property while idle if such services are necessary to preserve the buildings used in the mine or mining equipment required to work the property, *provided* it is intended to use these things again within a reasonable time.

6.　Unwatering a mine if done for the purpose of resuming operations, but not if done in order to permit someone to examine the property.

7.　Driving a cross-cut tunnel for the purpose of cutting veins or other types of deposits on a claim or group of claims, even though said tunnel is not started on the claim or claims that it is desired to hold and even if said tunnel has not yet been driven far enough to pass beneath said claim or claims.

Expenditures That are not Allowable as Annual Labor: Court decisions (which likewise might be reversed by other courts) have been made to the effect that the cost of doing certain things on or in connection with lode claims do *not* fulfill the annual labor provisions, as follows:

1. Expending money in travel in an endeavor to finance mining operations or to procure equipment.

2. Placing on the claims tools, implements, lumber, and other things that are not used to any extent and are subsequently removed.

3. Erecting on the claim a log cabin to be used by laborers, or a combined residence and blacksmith's shop, but other decisions have approved the erection of necessary buildings such as bunk houses, boarding houses, stables and blacksmiths' shops as satisfying the requirements of the annual labor law, so expenditures of money for erecting buildings other than shaft houses and structures used directly in connection with mining operations involve some degree of risk. It should be said, however, that the courts are inclined to be liberal where good faith is shown and buildings are obviously used in connection with mining operations.

4. Removing small quantities of rock or ore from workings from time to time and having it assayed or tested in any way.

5. Sharpening picks or drills when it is not proven they were used on the claim.

6. Erecting a stamp or other type of ore mill on or off the claim, although intended to be operated on ore exclusively obtained from the claim, is not a valid form of assessment work since it does not facilitate the *mining* of ore from the claim even though the existence of such a mill is necessary to make profitable mining of the ore possible.

7. Erecting a smelter (for the reason stated in paragraph 6).

8. Erecting a lime kiln for the purpose of making lime to be used in the construction of mine structures.

Failure of Partner or Partners to Contribute Towards Assessment Work: The Federal statutes provide that: "Upon the failure of any one of several co-owners to contribute his proportion of the required expenditures, the co-owners, who have performed the labor or made the improvements as required, may, at the expiration of the year, give such delinquent co-owner personal notice in writing, or notice by publication in the newspaper published nearest the claim for at least once a week for ninety days; and if, upon the expiration of ninety days after such notice in writing, or upon the expiration of one hundred and eighty days after the first newspaper publication of notice, the delinquent co-owner shall have failed to contribute his proportion to meet such expenditures or improvements, his interest in the claim by law passes to his co-owners who have made the expenditures or improvements as aforesaid. Where a claimant alleges ownership of a forfeited interest under the foregoing provision, the sworn statement of the publisher as to the facts of publication, giving dates, and a printed copy of the notice published, should be furnished and the claimant must swear that the delinquent co-owner failed to contribute his proper proportion within the period fixed by the statute."

The Arizona statutes further provide that: "Whenever a co-owner shall give to a delinquent co-owner the notice in writing or notice by publication to contribute his proportion of the expense of annual labor as provided by the laws of the United States, an affidavit of the person giving such notice, stating the time, place, manner of service, and by whom and upon whom such service was made, shall be attached to a true copy of such notice, and such notice and affidavit recorded. Ninety days after giving the notice, or, if such notice is given by publication in a newspaper, there shall be attached to a printed copy of such notice an affidavit of the editor or publisher of such paper, stating the date of each insertion of such notice therein, and when and where the newspaper was published during that time, and such affidavit and notice shall be recorded one hundred and eighty days after the first publication. The original of such notice and affidavit, or the records thereof, shall be *prima facie* evidence that the said delinquent has failed or refused to contribute his proportion of the expenditure, and of the service or publication of said notice; unless the writing or affidavit provided for in the following section is of record.

"If such delinquent shall, within the time required by the laws of the United States, contribute his proportion of such expenditures, such co-owner shall sign and deliver to the delinquent, a writing stating that the delinquent, by name, has, within the time required, contributed his share for the year upon the claim, and further stating therein the district, county, and state wherein the claim is situate, and the book and page where the location notice is recorded. Such writing shall be recorded. If such co-owner fail to sign and deliver such writing to the delinquent within twenty days after such contribution, he shall be liable to the delinquent for a penalty of one hundred dollars, and the delinquent, with two disinterested persons having personal knowledge of said contribution, may make an affidavit, setting forth the manner, amount, to whom and upon what mine such contribution was made. Such affidavit may be recorded, and shall be *prima facie* evidence of such contribution."

TUNNEL SITES

The Arizona statutes contain no reference to the location and retention of tunnel sites so the Federal statutes and regulations alone apply in this state. Unfortunately, they are vague and court decisions made under them are contradictory. It is believed, however, that the following statements accord with approved practice.

The maximum length permitted is 3,000 feet.

The location of a tunnel site is accomplished by posting a location notice at the mouth of the tunnel which is considered to be the point at which it will first enter cover. There follows an example of an adequate tunnel site location notice. The words or figures in italics will, of course, be changed to meet existing conditions:

"The *Gold Spot Tunnel and Tunnel Sites* located this *first day of August, 1934* by *A. D. Jones.* Course *S. 45° W., 3,000 feet.* *Squaw Peak* bears *S. 10° E.* and *Mineral Monument No. 14* bears *N. 42° 15' W.* from the mouth of the proposed tunnel where this notice is posted.

"The height of the proposed tunnel is 7 *feet* and the width 5 *feet.*

"I claim all lodes that may be discovered in this tunnel, not previously located, within a distance of 1500 feet on each side of the center line of this tunnel. I also claim for dumping purposes, at the mouth of said tunnel, a square tract of land *250 feet* on each side as staked upon the ground."

A. D. Jones.

Within ninety days after the date of the location, the following things must be done:

1. Mark the line of the tunnel on the surface of the ground with two parallel lines of substantial posts that project at least 4 feet above the surface, or by substantial stone monuments at least 3 feet high, arranged in pairs. Each pair of posts or monuments must stand directly above opposite sides of the tunnel. In other words, if the width of the proposed tunnel is to be 5 feet, each pair of posts or monuments must be 5 feet apart. Each pair of posts or monuments must be so spaced that at least one pair is visible from any point along the line of the proposed tunnel.

Some authorities claim that it is necessary to set only one line of monuments above the center line of a proposed tunnel, but Land Office rules specify that the *lines* of the tunnel must be marked on the ground so it is safest to set monuments above each side of the tunnel.

If wooden posts are used, the name of the tunnel and the number of the post should be cut on each of them, while, if stone monuments are erected, the same information on slips of paper should be enclosed in tin cans and built into each monument.

2. File a copy of the location notice, attested before a notary, with the County Recorder. As is true with the location notices of lode claims, the location of a tunnel site need not be an exact duplicate of the location certificate that is filed with the County Recorder, but that certificate must contain all the information given on the location notice and must be attested before a notary. The description of the tunnel site and the dump site, if one is included in the tunnel site location, should be given by bearings and distances on the recorded certificate.

The tunnel must be started within six months after the date of the location, and failure to prosecute the driving of the tunnel for a period of six months at any time constitutes an abandonment of all rights to blind lodes that have been cut by the tunnel, which have not been formally located. Such cessation of work does not prevent, however, the tunnel locator from resuming work later and acquiring title to other blind lodes that may be subsequently cut by the tunnel.

If previously unlocated lodes are cut in the tunnel, the tunnel owner has the right to locate a maximum of 1,500 feet along each lode. The claims thus acquired may extend 1,500 feet to one side of the tunnel or part of each 1,500-foot claim may be on one side and part on the other side of the tunnel. All of the things outlined as necessary to be done when the discovery of a lode is made on the surface must be done when the lode is discovered in a tunnel except that no discovery shaft or cut need be dug. The tunnel work that led to the discovery takes the place of such a shaft or cut. The location notice should be posted on the surface at approximately the point where the lode would reach the surface if it extended upward with the dip observed at the point discovered in the tunnel, and the location notice must, of course, be posted midway between the side lines of the claim located. It should also, probably, be posted directly above the lines of the tunnel. The location notice must state that the discovery was made in the tunnel, and the number of feet from the mouth of the tunnel to the point of discovery must be mentioned.

There are good grounds to believe that it is not necessary to post a location notice or monument the claim on the surface, and that it will suffice to post the location notice at the mouth of the tunnel and file a copy with the County Recorder, but it is probably safer to monument the claim and post the notice of the location on the surface. Certainly the claim must be monumented on the surface before it can be patented.

The location of a tunnel site does not give the locator a right-of-way beneath previously located claims.

If someone locates a lode along the line of a tunnel *after* the tunnel site was located and the lode is subsequently cut by the tunnel, the surface location becomes invalid.

No assessment work is required on tunnel sites and they cannot be patented.

MILL SITES

The laws and regulations relating to mill sites are fairly clear and lacking in ambiguity so there is little doubt of the correctness of any of the following statements.

Size: Mill sites cannot exceed 5 acres in area. Rectangles that measure 330 feet by 660 feet or 400 feet by 544.5 feet or squares that measure 466.69 feet on a side will enclose 5 acres.

Must be on Non-mineral Land not Adjacent to Lode Claims: The Federal statutes provide that only non-mineral land can be located or patented as a mill site. If a patent is desired, the non-mineral character of the ground must be established by the affidavits of disinterested witnesses. The Federal statutes also provide that a mill site cannot be adjacent or contiguous to a claim or group of claims owned by the locator of the mill site.

Need Not be Named: It is not necessary to give a name to a mill site, but it is desirable to do so, and it will be found convenient to refer to it by name.

How Located: 1.   Post a location notice at one of the corners. This notice may be similar to that used for a lode claim and must contain all the data required on such a notice excepting for the small changes necessitated by the fact that no lode is involved.

2.   Within ninety days, monument the mill site by setting posts or stone monuments of the sizes specified for lode claims at each corner of the mill site. Place the name of the mill site and the number of the corner on each post or in a tin can built into each stone monument.

3.   Within ninety days, record the location certificate, attested before a notary, with the Recorder of the county in which the mill site is located. This location certificate may be more elaborate than the location notice and it will be desirable to give a description of the boundaries of the millsite by bearings and distances. Nothing given on the location notice must be omitted on the location certificate.

Two Classes of Mill Sites: Although claims located as just set forth are called mill sites, that term is not affixed to them by the Federal or State statutes. In fact, they are not mentioned in the State Mining Code. The Federal statutes merely describe them as claims which contain no lodes but are "occupied for mining or milling purposes." There are, therefore, two classes of mill sites: (a) Mill sites occupied for mining purposes, and (b) mill sites occupied for milling and smelting purposes.

Congress has allowed each owner of one or more lode claims to occupy 5 acres of open, non-mineral land on the theory that he may require such additional ground for bunk houses, cabins, storage buildings, etc., at some point more desirably located than on the mineralized ground owned by him where the actual mining operations are to be conducted. Such a tract of land is not really a mill site, but it is so styled.

If the mill site is actually to have some form of reduction works erected thereon, it is appropriately named, and it is not necessary that the locator of such a mill site own any lode mining claims.

Retaining Mill Sites: In order to retain a mill site, the location must be followed by actual occupancy or the erection of substantial and valid improvements such as:

1. Erecting a pumping plant to elevate water to a mine.
2. Constructing a reservoir in which to store water for use at an operating mine.
3. Storing ore from an operating mine.
4. Erecting and using housing facilities for the owner and workmen, storage buildings, shops, etc.
5. Erecting any type of reduction works.

No Assessment Work Required on Mill Sites: It is not necessary to spend any particular amount of money on a mill site in order to retain it. Actual occupancy or the presence of substantial and valid improvements on the mill site is all that is required, as has been stated.

Patenting Mill Sites: Mill sites may be patented just like lode claims and the price paid to the government for the land is $5.00 per acre. If the mill site is owned by the owner of a neighboring, but non-contiguous, claim, the mill site may be patented with the claim no matter how little money has been spent on the mill site. In other words, the statute requirements that $500 must have been spent in improving each lode claim before it can be patented does not apply to mill sites. It is essential, however, that the mill site be actually used in connection with the mining or milling of ore from the lode claim with which it is patented. A mill site may also be patented separately *after* a patent has been obtained on the lode claim or claims operated in connection with it. Finally, a mill site owned by someone who owns no lode claims in the vicinity may also be patented, but, in that case, some form of reduction works must have been erected thereon before a patent will be issued.

THE APEX LAW OR EXTRALATERAL RIGHTS

In the United States, the owners of all types of real estate except lode mining claims own everything included within vertical planes passed downward through the boundaries clear to the center of the earth unless the mineral rights are excluded in the deeds. The owner of a lode mining claim, however, whether held by right of location or by patent, can follow each lode that outcrops or has it apex on the claim downward as far as he wishes to go, no matter how far that may be beyond a vertical plane passed through the side line of claim on which the lode outcrops. He cannot, however, trespass beyond the vertical planes passed through the parallel end lines or on the surface of mining claims held by others under which his lode may pass. The Federal statutes governing this matter read as follows:

"The locators of all mining locations ... shall have the exclusive right of possession and enjoyment of all the surface included within the lines of their locations, and of all veins, lodes, and ledges throughout their entire depth, the top or apex of which lies inside of such surface lines extended downward vertically, although such veins, lodes, or ledges may so far depart from a perpendicular in their course downward as to extend outside the vertical side lines of such surface locations. But their right of possession to such outside parts of such veins or ledges shall be confined to such portions thereof as lie between vertical planes drawn downward, as above described, through the end lines of their locations, so continued in their own direction that such planes will intersect such exterior parts of such veins or ledges. And nothing in this section shall authorize the locator or possessor of a vein or lode which extends in its downward course beyond the vertical lines of his claim to enter upon the surface of a claim owned or possessed by another."

OWNERSHIP OF PLACER DEPOSITS ON LODE CLAIMS

The owner of a lode claim is also the owner of any placer deposits that may exist thereon, which were not located prior to the time when the lode claim location notice was posted. It is, however, essential that the lode claim should have been located legally on a deposit of mineral in place, in vein or lode form, and that all of the requirements of the statutes that govern the acquirement and holding of lode claims should have been met.

OWNERSHIP OF LODES WITHIN THE BOUNDARIES OF PLACER CLAIMS

A placer location does not cover any lodes that may exist within it. Either the holder of the placer claim or any stranger may locate and hold a lode included within an unpatented placer claim by following the procedure laid down in the Federal and State statutes for locating lode claims, just as though no placer claim had been located around the lode. Furthermore, strangers are permited to enter peaceably upon an unpatented placer claim, without the owner's consent, to locate a lode claim. After a placer claim has been patented, its owner also owns any lodes included within it if their existence were unknown when the patent was issued.

SELLING AND TAXING LODE CLAIMS

A lode claim held by right of location and compliance with the Federal statutes that relate to the performance of annual labor may be sold or leased like any other real estate. Unpatented lode claims cannot be taxed as real estate, however, although buildings and their contents placed on them may be taxed.

PATENTING LODE CLAIMS

A lode mining claim located on the public domain may be patented or purchased from the United States government, after at least $500 worth of work has been done upon it, for $5.00 per acre or fraction of an acre, plus various fees.

If the claim that it is desired to patent is on unsurveyed land, the claimant must also pay to have the claim surveyed by a United States Mineral Surveyor. Patent procedure is so complicated that it would serve no good purpose to outline it in this bulletin, and anyone who contemplates applying for a patent should consult an attorney-at-law.

PART IV

SOME HINTS ON PROSPECTING FOR GOLD

By G. M. Butler,

Director, Arizona Bureau of Mines

Unfavorable Areas.

The saying that "gold is where you find it" is certainly true. There are, nevertheless, certain conditions that are so unfavorable to the occurrence of gold in any considerable quantity that prospectors would do well to avoid areas in which these conditions exist. Among such unfavorable areas are the following:

1. Areas where large masses of granite and related, coarse grained, crystalline igneous (once molten) rocks outcrop, particularly if these outcrops are not cut by dikes or other intrusions of finer grained, usually light colored igneous rocks such as porphyry, rhyolite, or andesite.

2. Areas where large masses of gneisses and the other crystalline schists outcrop unless they are cut by or in the vicinity of dikes or other intrusions of igneous rocks.

3. Areas where large masses of sedimentary rocks such as limestone, sandstone, and shale outcrop unless they are cut by dikes or other intrusions of the relatively fine grained, light colored igneous rocks previously mentioned, and, even where so cut, sedimentary areas rarely contain workable quantities of gold unless the sediments have been metamorphosed (changed in character by pressure and heat) to marble, quartzite, or slate.

4. Areas where large masses of dark colored, relatively heavy igneous rocks, such as peridotite, diabase, and basalt or malpais outcrop.

5. Areas in which nothing but the unconsolidated or loosely consolidated material that fills the valleys between the mountain ranges in southern and western Arizona outcrops.

A study of the geological map of Arizona described in the back of this bulletin will enable anyone to learn where the great sedimentary and valley-fill areas are situated in Arizona and to avoid them. The great granitic and schistose masses are also shown on that map, but its scale is so small that intrusions of igneous rocks hundreds of feet wide cannot be shown thereon. It is, therefore, necessary to examine the granitic and schistose areas to determine whether the conditions there are unfavorable.

It is not true that valuable gold lodes never occur in areas described as generally unfavorable, such as in a great mass of granite without intrusions of other igneous rocks, for instance, but a prospector will usually save time and money by avoiding such areas.

Favorable Areas.

1. Probably the most favorable area in which to prospect for gold is one where the country rock is made up of surface flows, sills, dikes, and other intrusions of relatively fine grained, light colored, Tertiary igneous rocks such as rhyolite, trachyte, latite, phonolite, and andesite.

2. As has already been suggested, prospecting in areas where there are outcrops of granitic or schistose rocks that are cut by dikes or other intrusions of relatively fine grained, light colored igneous rocks may prove profitable.

3. Areas in which the country rock is some type of porphyry, especially if several varieties formed at different time are found there, may contain deposits of gold that can be worked profitably.

4. Gold lodes that may be worked profitably are sometimes formed around the borders of great masses of granitic igneous rocks, both in the granitic and in the surrounding rocks, but more commonly the latter.

5. Areas in which some gold has already been found are naturally more favorable than places that have never produced any gold. This statement applies particularly to areas where considerable prospecting has been done—any area in continental United States excepting much of Alaska.

Structures that May Contain Gold.

The term "lode" as used in the Federal statutes is applied to all deposits of "mineral in place" formed beneath the surface of the earth. When gold is found in sand or gravel, either loose or cemented, the occurrence constitutes a gold placer deposit.

The gold found in lodes has arisen from great depths in solution and has been precipitated or deposited from such solutions by relief of pressure, cooling of the solutions, and other causes. There must, then, be some form of opening or zone of weakness through the rocks along which the solutions may rise. Although the mineral-bearing solutions sometimes find their way toward the surface through masses of porous material that have roughly the form of vertical cylinders, lodes are usually long and relatively narrow.

From what has been said, it should be evident that the existence of openings or lines of weakness in the rocks is the condition that fixes the position of a lode.

Most frequently, the solutions rise through a crack or fissure that extends a long way downward or through a series of interlacing cracks or fissures.

If the solutions precipitate or deposit ore minerals and gangue (the worthless minerals deposited along with ore minerals) in a single, clean-cut fissure, filling it full of ore minerals and gangue, the result is a simple fissure vein. If there has been movement of one wall relative to the other, parallel to the fissure, and ore minerals and gangue have been deposited in the fissure, the result is a fault fissure vein.

Both simple and fault fissure veins are commonly called by miners "true fissure veins."

Where movement or faulting has occurred at considerable depth and the pressure is great, the walls of the fissure may be polished or "slickened" and clay-like, finely crushed rock called "gouge" may be formed. Subsequent deposition of mineral in the fissure would force the walls apart and the gouge may then be found along both walls. After the vein has formed, the movement or faulting may be resumed and "slickensides" and gouge may then be formed in the vein material itself, or, if the pressure is then relatively light, the vein material may be broken into fragments, thus forming one type of so-called "brecciated vein."

If the mineral-bearing solutions rise through and deposit ore minerals and gangue in the zone of interlacing, closely spaced cracks, the result is a shear zone and that is a very common type of lode.

The contact between two different rocks, especially an igneous rock and something else, as porphyry and schist, or two different rocks, is usually a line of weakness. If a mass of igneous rock is involved, shrinkage occurs on cooling and that shrinkage tends to cause the igneous mass to pull away from the rocks with which it is in contact. If solutions pass up through and deposit ore minerals and gangue along such a line of weakness, the result is called a contact fissure vein.

There are a number of other varieties of lodes, but they are rarer than the types mentioned and it would unduly lengthen this chapter to describe them.

It should be emphasized that most of the structures mentioned contain so little valuable material that they cannot be mined profitably. Nevertheless, they are the things that must be sought and any such structure found should be carefully investigated. Furthermore, this investigation should not be confined to the vein matter itself, but should include the walls since they are sometimes impregnated with material that has been precipitated from solutions that have penetrated them. They should be assayed, especially if they appear to be softened and altered. Moreover, the wall rock itself is sometimes completely replaced with vein material to considerable distances from the vein itself.

Surface Characteristics of Gold Lodes.

The actual outcrop of a lode usually consists of a zone or band of material that differs in character from the material on both sides of it, but the discovery of such a zone or band does not prove that a lode has been found. It might consist, for instance, of a bed of limestone between beds of shale, all turned up at a steep angle and eroded. The outcrop of a lode has, fortunately, other characteristics that make it possible to recognize it as follows:

1. It does not consist entirely of any single species of rock. It may be composed almost entirely of some species of rock, but there will be veinlets running through it in that case.

2. Because the solutions that form the lode usually deposit sulphide of iron along with the gold, and sulphide of iron is changed to yellowish brown to dark brown oxide of iron when exposed to the atmosphere, the outcrops of lodes are usually heavily "iron stained." Stains produced by the oxidation of minerals that contain other metals than iron may also be present, such as green or blue (copper), black (manganese), light yellow (molybdenum or lead), lilac (cobalt), etc.

3. Because some of the iron and all of the sulphur, as well as other soluble ore and gangue minerals, may be carried away in solution when the lode is exposed to atmospheric weathering, the outcrop is apt to be decidedly porous. The grade of dense outcrops is usually very low.

4. The gangue mineral most commonly associated with gold is quartz and a gold lode that does not contain some quartz is rare indeed. Low grade gold veins that consist almost entirely of quartz are not uncommon. Other gangue minerals sometimes associated with gold (usually one or two of them are present except in the almost pure quartz or quartz-pyrite veins) are carbonates (particularly calcite), adularia feldspar, sericite (very finely granular white mica), fluorite, etc.

It should not be assumed that a valuable discovery has certainly been made when a lode that consists of heavily iron-stained, porous rotten quartz, where it outcrops, has been found. Such material often contains little or no gold. By crushing and panning it, the presence of gold can sometimes be ascertained, but the only safe method to use to learn whether it is valuable is to have it assayed by a reputable assayer.

It should be remembered, however, that few lodes contain even approximately the same proportion of gold for any distance along their outcrops. Gold, as well as other ore minerals, is concentrated at certain points in the lode in what are called "shoots," "ore shoots," "ore channels," etc., of relatively limited extent. Moreover, values may be concentrated along one wall or at a certain distance from a wall. If, then, a lode has been discovered, samples at numerous points along its outcrop and from various positions in it should be crushed and panned or assayed before deciding whether the ore is or is not valuable.

Identifying Gold.

Sometimes specks, grains, or thin plates of gold are visible in the outcrop of a lode and, in that case, they may be recovered by crushing and panning. It is, therefore, important that a prospector should be able to identify such material and, especially, to distinguish it from other substances with which it may be confused.

Ores of nearly all substances excepting the very rare metals like gold, silver, and platinum, usually occur in masses of such size that they may easily be seen, and anyone familiar with these minerals will recognize them by their physical characteristics. It is usually unnecessary for a mineralogist to assay a specimen in order to ascertain whether it contains considerable quantities of copper, zinc, lead, manganese, etc., for instance. Such is not the case with gold, however. In the great majority of instances, it occurs in tiny grains that are distributed through the gangue or included within other minerals and are quite invisible. They may be sometimes recovered by panning, but frequently they are so small or are so firmly locked up in other minerals that only fire assaying will reveal their presence. Rarely, gold is combined with tellurium to form minerals that are called tellurides and they can be recognized, if present in visible grains, by a mineralogist who is familiar with their characteristics, but there is no known deposit of telluride gold ore in Arizona, and it is not likely that one will be found. It is fortunate that telluride ores do not occur commonly here since special methods of assaying and treating them must be used.

In spite of the fact that visible gold is not common in lodes, it is important, as stated, that prospectors should be able to recognize it when it is found. They should be able to do so by the fact that it is the only soft, yellow, metallic lustered substance found in nature that may be easily flattened without breaking and easily cut with a knife blade or indented with a needle or any other small, sharp-pointed instrument. It may be confused with pyrite, chalcopyrite, and several other sulphide minerals, but pyrite is so hard that it cannot be scratched and all sulphide minerals that resemble it crush into a black powder instead of cutting cleanly. Gold is sometimes confused with plates of yellow mica, but they are much softer and yield a white powder when scratched with the point of a needle.

Sometimes there is so much silver present as an impurity in the gold that the color becomes almost silver-white, but it is rarely that there is not some yellow tint to the alloy. Chemical tests are necessary to distinguish gold that contains 20 per cent or more silver from metallic silver.

Seeking Gold Lodes.

Many gold lodes, but by no means all of them, when exposed to erosion, disintegrate and the gold in them washes down into the beds of streams that drain the area. A common method of prospecting for gold lodes is to pan or use a small dry washer on the material from washes and stream beds and to endeavor to trace the placer gold back to its source. If placer gold is irregular in shape and more or less jagged or if quartz is still embedded in the grains or nuggets, it has not moved far from its source. If, however, it is smoothly rounded or flattened and contains no quartz, its source may be many miles away.

If particles of gold have been found in a stream bed or wash and have been traced up stream to a point where they suddenly become much less numerous or disappear, the gold has probably come from a lode directly up hill from the point where it was last found in the stream bed or wash. By trenching or panning and dry washing the earth on both sides of the wash and up hill therefrom, it may be possible to determine in which direction to seek the source of the gold. If the direction from which it came can be thus ascertained, it then becomes necessary to re-member what has been said about the appearance of the out-crop of a lode and to hunt for one.

It should be said, however, that even if the outcrop of a lode that has been the source of placer gold is found, it may be too small or two low grade to be mined profitably, although the pla-cer deposit formed of the gold released from it is large and rich. It is always possible that this placer gold may have come from a wider or a richer part of the lode, that has been removed by erosion.

It is also true that some valuable gold lodes do not yield much, or any, placer gold, and this statement may apply to all the lodes in an entire district, such as the Oatman district in Arizona. The absence of placer gold should not, then, deter a prospector if other conditions appear not unfavorable and, especially, if gold has been found in the district.

When no placer gold is found in a district, it is necessary to search for a lode or indications of the existence of a lode without the guidance of placer gold particles that have been shed from it. One should, then, search through the gulches, washes, and stream beds for pieces of "float" (fragments of ore minerals and gangue, such as "rusty quartz," broken from a lode and washed down therefrom). Their angularity and smoothness will give some in-dication of the distance they have travelled, but they are usually too scarce to be followed back to their source in the same way that placer gold can be traced.

Having decided to try to find a lode on a given hill or a mountain, it is usually best to walk along the slope, endeavoring to keep on the same level, and search carefully for float or other indications of the existence of a lode. If nothing is found at one level, climb a few hundred feet and again circle the hill or mountain at that level. Continue alternately to climb and circle until the top has been reached. While climbing or circling, if the ground is covered with broken rock or soil, look for the following things that may be caused by the existence of a lode at the point where they are seen:

1. A trench or ditch that does not run directly down the slope of the hill or mountain.

2. A sudden change of slope.

3. A sharp notch that crosses a ridge that has a rather uniform altitude on both sides of the notch.

4. Several springs in a line.

5. A sudden change in the kind or quantity of vegetation (may indicate a contact or, if the change in vegetation is found over a narrow strip of ground, a lode may be beneath).

6. A change in the nature of the rock fragments (indicates a contact).

It is true that there are many possible causes for the existence of all the things just mentioned, and that the presence of a lode is only one of them, and not a very likely one. Nevertheless, some trenching should be done when such features occur, especially if float has been found below, and not above. When digging in search for a lode on a hill side that is covered with several feet of loose rock and soil, it should be remembered that float works down hill and a lode should be sought several feet above the point where the highest float is found.

Sampling Lodes.

When a lode has been found and it is decided to have one or several assays made, most inexperienced prospectors make the serious mistake of collecting samples that are altogether too small. The majority of the samples sent to the Arizona Bureau of Mines to be assayed are each a single fragment weighing a pound or less or two or three smaller fragments. Such "specimen" assays are a complete waste of time and money. They furnish almost no indication as to whether a lode has been found that can or cannot be mined profitably. Not only is valuable ore concentrated in shoots (Page 249), but the grade often varies greatly at distances of only a few inches in a shoot. A single specimen might easily assay $50 a ton and yet represent only 2 inches of ore. On the other hand, barren specimens can often be found in lodes that will run $50 a ton, for instance, as a whole.

The only safe way to do is to cut very large samples clear across a lode so that each truly represents the ore that might be mined at that point, and crush and quarter them down or send the whole sample to an assayer. He will charge no more for assaying a 10-, 20-, or 30-pound sample than for a one-pound sample.

INDEX

F

G

H

McMillen, G. W. ... 136
McReynolds, T. C., Jr ...166
Magma Copper Co. ... 169
Mammon mine ... 175
Mammoth mine, Goldfields district ... 167
Mammoth or Hubbard mine, Yavapai County ... 27
Mammoth mine, Mammoth district ...170-174
Mammoth prospect, Yuma County ... 136
Mandalay property ... 108
Manning, W. H. ... 126
Manzoro Gold Mining Co. ... 121
Margarita mine ... 191
Maricopa County ... 156
Maricopa mine ... 164
Mariquita prospect ... 136
Martin, F. J. ... 155
Martinez district ... 69
Mary E. vein ... 109
Mascot property ... 77
Max Delta mine ... 166
Mayhew, Felix ...140, 141
Merrill property ... 192
Mesothermal veins ... 15
Mesozoic veins ...15, 22
Metcalf vicinity ... 187
Mex Hill property ... 164
Midnight mine ... 96
Midnight Test mine ... 32
Milevore Copper Co. ... 64
Mills, E. ... 146
Mill sites, locating of ... 234
Mineral Park vicinity ... 112
Mine shafts ... 207
Minnehaha vicinity ... 59
Minton, D. C. ... 53
Misfires ... 205
Mitchell, H. C. ... 51
Mitchell vein ... 96
Mitchell, C. W. ... 162
Mocking Bird mine ... 78
Mohave County ... 73
Mohave Gold Mining Co. ... 96
Mohave Gold property ... 79
Mohave Mountains ... 115
Mohawk mine, Mammoth district ...170-174
Mohawk property, Bradshaw district ... 55
Molybdenum Corp. of America ... 171
Monarch mine, Cherry Creek district ... 29
Monarch mine, Oro Blanco district ... 191
Monarch property, Union Pass district ... 108
Money Metals mine ... 40
Monte Cristo property ... 33

Mooney, F. R. ... 181
Moore, R. W. ... 92
Morenci region ... 185
Morgan mine ... 179
Mormon Girl mine ... 164
Morrison, G. ... 185
Mossback mine ... 98
Moss mine ... 94
Mount Union mine ... 50
Mudhole mine ... 35
Music Mountain district ... 108
Myers, G. ... 133

N

Naething, F. S. ... 173
National Gold Corporation ... 32
Neal, W. ... 133
Nelson, C. W. ... 66
Never-get-left property ... 77
Nevius, J. N. ... 68
New Comstock Mining Co. ... 103, 106
New United Verde Copper Co. ... 32
New Year property ...171-174
Nigger Brown mine ... 51
Nielson, F. ... 115
Niemann, E. F. ...103, 105
North American Exploration Company ... 103
Northern Black Mountains ... 78
North Star mine ... 139
Nixon, S. ... 66

O

Oatman Associates Mining Co. ... 92
Oatman district ... 80
O'Brien mine ... 63
Occidental vein ... 40
Octave mine ... 66
O. K. mine ... 77
Old Glory mine ... 189
Old Hat district ... 170
Old Maid mine ... 135
Old Terrible mine ... 121
Olla Oatman shaft ... 90
Operations at small gold mines ... 195
Orem, C. L. ... 60
Orion mine ... 95
Oro Belle mine ... 59
Oro Blanco district ... 187
Oro Blanco mine ...188, 190
Oro Flame mine ... 40
Oro Grande mine, near Wickenburg ... 62
Oro Grande Mining Co. ... 42
Oro Plata mine ... 114
Osborne, E. S. ... 127
Ox Bow mine ... 184